T3-BFD-434

Alexander Lipski

Thomas Merton and Asia: His Quest for Utopia

CISTERCIAN STUDIES SERIES: NUMBER SEVENTY-FOUR

Thomas Merton and Asia: His Quest for Utopia

by

Alexander Lipski

Cistercian Publications
Kalamazoo, Michigan
1983

Cistercian Publications wishes to express its appreciation to the following publishers for permission to quote passages in this book. George Allen & Unwin, Ltd., Hemel Hempstead; Costello Publishing Co. Inc., Northport, New York; Doubleday & Company Inc., New York; Lover Publications, Inc., New York; Farrar, Straus & Giroux, Inc., New York; Harcourt Brace Jovanovich, Inc., New York; Harper & Row, Publishers, Inc., New York; New Directions Publishing Corporation, New York; Pantheon Books, New York; Philosophical Library, Inc., New York; Routledge & Kegan Paul Ltd., London; Robinson & Watkins, London; and The Merton Legacy Trust.

Typeset by Gale Akins, Kalamazoo
Printed in the United States of America

Library of Congress Cataloging in Publication Data
Lipski, Alexander, 1919–
Thomas Merton and Asia.

(Cistercian studies series ; no. 74)
Bibliography: p. 87
1. Merton, Thomas, 1915–1968. 2. Asia—Description and travel—1951– 3. Asia—Religion. I. Title. II. Series.
BX4705.M542L56 1983 271'.125'024 83-7516
ISBN 0-87907-874-X
ISBN 0-87907-974-6 (pbk.)

To my wife
Ruth–Maria

TABLE OF CONTENTS

PREFACE

THAT THOMAS MERTON died in Asia was certainly not fortuitous. His abbot, Flavian Burns, in his homily at the Eucharist the day after his death, commented: 'He even saw a certain fittingness in dying over there amidst those Asian monks, who symbolized for him man's ancient and perennial desire for the deep things of God'.[1] Well in advance of Vatican II, Merton recognized the need to dialogue with representatives of Asian spirituality in the conviction that Asian wisdom could enrich his own Catholic tradition and contribute to a renewal of Christianity, especially of monasticism. It was primarily the ancient monastic and contemplative tradition of the East which attracted him and which he hoped to use as an antidote to a crudely materialistic, manipulative, over-intellectualized and egocentric West. Studying the spiritual treasures of Asia, Merton discovered striking similarities with the teachings of Christianity—teachings which modern western society had seen fit to discard at its peril. He was dismayed to find that the West had not only forgotten much of its own precious spiritual heritage, but had imposed its modern, so-called progressive value system on Asia. Merton eventually became a crusader for a return to a life balanced between contemplation and action, a life of self-transcendence rather than egocentricity. Above all, he became a fervent advocate of cooperation between East and West. Asian wisdom undoubtedly enlarged Merton's vision and deepened his insight into life's mystery. Greatly enriched, he was able to kindle among many of his contemporaries an interest in Asian spirituality, and to promote a dialogue between Christian and Buddhist monks, thereby contributing to the essential task of bridging the gulf between Orient and Occident.

While there have been detailed studies devoted to Merton and Zen Buddhism, there has not so far been a study dealing with Merton's Asian interest in general. In particular, the role of Hinduism in the shaping of Merton's *Weltanschauung* has not been analyzed in depth. As we do this, we shall also attempt to establish the extent to which Merton's idealistic presuppositions colored his image of Asia and his interpretation of Asian religions, and thereby blurred his view of Asian reality.

I wish to express my gratitude first of all to Sister Mary Luke Tobin SL, a close friend of Merton, who encouraged me to undertake this study. I should like to express my deep appreciation as well to all those who rendered assistance at the Thomas Merton Studies Center at Bellarmine College. Its Curator and Associate Director, Dr Robert Daggy, helped me in numerous ways with my research and made my stay at the Center a most worthwhile experience. Many thanks are due to the Thomas Merton Legacy Trust for their generous cooperation with this venture. I owe a special debt to Brother Patrick Hart who arranged for me to stay at Gethsemani Abbey and who was kind enough to share some of his insights derived from many years of acquaintance with Thomas Merton. Cistercian Publications, in particular the Editorial Director, Dr Rozanne Elder, deserve special mention for providing constructive criticism and invaluable suggestions. Last but not least I wish to thank my wife, Ruth Maria, for her support throughout my work on Merton.

Chapter I
MERTON AND ASIA AFAR

DISCONTENT WITH HIS OWN, modern western, society Thomas Merton throughout his life sought a more perfect society either in the past, the European Middle Ages, or in the non-western world, Asia. In 1925, at the age of ten, Merton was taken by his father to France, the country of his birth, where he was 'to live and drink from the fountains of the Middle Ages'.[1] Describing in retrospect his visit to the southern French town of St. Antonin, he romanticized:

> The whole landscape, unified by the church and its heavenward spire, seemed to say: this is the meaning of all things; we have been made for no other purpose than that men

> may use us in raising themselves to God, and in proclaiming the glory of God . . . . [2]

But his own age, his own society filled him with abhorrence. At various times he blamed, jointly or individually, human greed, selfishness, capitalism and technology for the failings of western society. Most often he diagnosed the illness of modern humankind as alienation from its true self. Seeing no meaning in the society around him, he was further shaken when at the age of fifteen, in 1930, he lost his father—he had already lost his mother nine years earlier. The fifteen-year-old Merton embarked upon a life of sensual pleasure while expressing his dismay at the emptiness of modern life:

> And so I became the complete twentieth-century man. I now belonged to the world in which I lived. I became a true citizen of my own disgusting century: the century of poison gas and atomic bombs. A man living on the doorsill of the Apocalypse, a man with veins full of poison, living in death.[3]

Two years later, while studying at Oakham, England, Merton became a Communist. What was more logical for a young, immature rebel against society than to embrace the simplistic slogans and solutions of Marxism? He could easily agree with Marx's condemnation of capitalistic vices. Gross materialism, social injustice and inhumane wars were obviously caused by capitalist greed. When he entered Columbia University in 1935, he encountered many a Communist sympathizer, and for a while attended Communist party meetings. But disillusionment with the U.S.S.R., the standard bearer of Communism, was not long in coming, and Merton concluded that, when all was said and done, the Soviet Union represented only another brand of materialism. In the latter part of his life he compared the Soviet Union and the United States to Gog and Magog of the Book of Revelation:

> Gog is a lover of power. Magog is absorbed in making money: their idols differ, and indeed their faces seem to be dead set against one another, but their madness is the same: they are the two faces of Janus looking inward, and

> dividing with critical fury the polluted sanctuary of dehumanized man . . . .[4]

While in retrospect he made sarcastic remarks about his flirtation with Communism, he continued to the end of his life to be attracted to some aspects of Marxism. He envisioned a 'pure' Marxism, a Marxism—not yet practiced in any existing society—combined with christian principles.

Seeking to identify the causes of the depravity of modern western society, he castigated the divorce of science and technology from morality which had resulted in genocide and atomic holocaust. Simultaneously he attacked modern consumerism, the manipulation of mankind by media in the interest of furthering the manufacture of gadgets which enslaved rather than liberated mankind.[5]

Although he looked for simplistic causes to explain his 'disgusting century', he intermittently realized that his own age was not radically different from other ages: 'Underneath, it was the same old story of greed and lust and self-love, of the three concupiscences bred in the rich, rotted undergrowth of what is technically called "the World" in every age, in every class'.[6] And he admitted: 'The climate of irrationality, confusion and violence which is characteristic of such times as ours is after all nothing new'.[7] It is likely that he temporarily found peace by reassuring himself: *'Plus ça change, plus c'est la même chose'* (The more things change, the more they are the same). When he became a Catholic and accepted the Church's teaching concerning Original Sin, he should have become more accepting of human weaknesses and the resultant imperfection of a society which is to attain perfection only at the end of time, in a transfigured cosmos. While theoretically he accepted these teachings, intense yearning for a perfect society exempt from sinfulness caused him to disregard essential flaws in the human makeup.

The non-western world, so Merton argued, had not yet sinned, with the exception of regions, such as China and Japan, which have been infected with the western bacillus.

Although in general Merton was horrified at the mere thought of nuclear holocaust, he was willing to consider the advantages of an atomic war between Gog and Magog. In such a war the two super-powers might conceivably destroy one another but leave the southern hemisphere in sufficiently good condition for survival. Merton speculated, probably at a time of deep depression:

> In this new situation it is conceivable that Indonesia, Latin America, Southern Africa and Australia may find themselves heirs to the opportunities and objectives which Gog and Magog shrugged off with such careless abandon . . . . Let us also hopefully assume the partial survival of India and of some Moslem populations in central and northern Africa . . . . It will mean that the cerebral and mechanistic cultures, those which have tended to live more and more by abstractions and to isolate themselves more and more from the natural world by rationalization will be succeeded by the sections of the human race which they oppressed and exploited without the slightest appreciation for or understanding of their human reality. Characteristic of these races is a totally different outlook on life, a spiritual outlook which is not abstract, but concrete, not pragmatic but hieratic, intuitive and affective rather than rationalistic and aggressive . . . . [8]

This faith in the purity and sinlessness of the non-western world was hardly original with Merton. Many a western student of Asia has juxtaposed a spiritual, intuitive East to a materialistic, rationalistic West. Guilt over the abuses perpetrated by western *conquistadores* is also not uncommon among western intellectuals and is an accompaniment to the sense of defeatism expressed in terms of the 'Decline of the West'. Rarely, however, has the northern hemisphere been pitted against the southern hemisphere or southern Africa, Indonesia, northern Africa, South Asia and South America been lumped together. The inclusion of the Republic of South Africa strikes one as most incongruous, especially in view of Merton's staunch opposition to racism. He probably had the Zulus and not the white inhabitants of South Africa in mind. Most importantly, he overlooked the fact that a hieratic, intuitive and affective civilization has never existed anywhere, and consequently there is no way of estimating how it might function.

While sympathetic and enthusiastic for the whole of the non-western world, Merton concentrated his chief attention and pinned his hope on Asia, explaining his special attraction by saying, 'Asia has long been the most thriving home of monastic vocation'.[9] At one time he went so far as to exclaim: 'I am much closer to Confucius and Lao Tzu than to my contemporaries in the United States . . . .'[10] At another time he confided: 'I am as much a Chinese Buddhist by temperament and spirit as I am a Christian . . . .'[11]

Merton's attraction to Asia developed gradually. The first concrete evidence of it dates back to November 1937, when he had come under the influence of Aldous Huxley. Since the 1930s Huxley, formerly a skeptic, had been attracted to mysticism and investigated Christian as well as Hindu and Buddhist mysticism. His newly acquired mystical views found expression in *Ends and Means* (1937), which Merton read at the suggestion of Robert Lax, a fellow student at Columbia University. Huxley not only aroused in Merton an interest in mysticism but also drew his attention to the resemblances in the experiences of eastern and western mystics. In particular, Huxley pointed out similarities in the views of the anonymous author of the *Cloud of Unknowing* and of Meister Eckhart with those of the Buddha and India's foremost philosopher, Sankara.[12] Merton as well as Huxley was drawn to the mystical emphasis of Asian religions, the importance placed upon contemplation and withdrawal from the world. Merton was deeply impressed by Huxley's contention that the practice of mystical techniques could bring about 'peace, toleration and charity',[13] ideals which he consistently cherished.

As an immediate consequence of reading *Ends and Means* Merton threw himself into the study of Oriental religious texts in translation by the Jesuit Leon Wieger. The first fruits of his encounter with the Orient were meager: ' . . . the strange jumble of myths and theories and moral aphorisms and elaborate parables made little or no real impression on my mind . . . . '[14] He later admitted that he had gone about his Oriental studies unsystematically and superficially. He had hastily concluded that eastern mysticism was not true mysticism, since it lacked the supernatural dimension. The value of meditation techniques appeared to him limited to muscular relaxation and suitable for inducing sleep through autosuggestion.[15]

The primarily negative impressions he formed of Oriental mysticism were partially neutralized by his encounter in June 1948 with an Indian yogi, Dr Mahanambrata Brahmachari. Dr Brahmachari's calmness and worshipfulness deeply impressed Merton. More importantly, Dr Brahmachari steered him towards Christian mysticism, urging him to read about the life of St Francis, St Augustine's *Confessions,* and Thomas à Kempis' *Imitation of Christ.*[16] Reading the classics of Catholic spiritually revealed to Merton that the West did have answers to ultimate questions. By means of contemplation western monks and nuns had attained inner peace and had risen above selfish greed and sense temptation to true freedom through a life in Christ. The lives of the saints reinforced an attraction to Catholicism which he had first vaguely felt while living in the medieval landscape of southern France. Thus the Hindu monk contributed towards his decision to become a Catholic in 1938 and his eventual entry into the Trappist Abbey of Gethsemani in 1941. That an Asian played a role in Merton's conversion to Roman Catholicism is a reflection of the ever increasing interplay between East and West which characterized Merton's pilgrimage through life and which in turn is symptomatic of the developing dialogue between Asian and Christian spiritual leaders. This connection is not unlike the influence that several Westerners, including Ruskin, Tolstoy and Sir Edwin

Arnold, had in strengthening Mahatma Gandhi in his Hindu faith.[17]

When Merton became a Trappist monk in 1941 he temporarily set aside his interest in Asia. He entered Gethsemani in the conviction that he could there best practice total surrender to God. By renouncing his ego, he hoped to find the rest which the world had not given him, especially the lunatic world which had become engulfed in World War II. 'By my monastic life and vows I am saying no to all concentration camps, aerial bombardments . . . . ' [18] And he exuberantly proclaimed:

> A monk is a man who has given up everything in order to possess everything. He is one who has abandoned desire in order to achieve the highest fulfilment of all desire. He has renounced his liberty in order to become free. He goes to war because he has found a kind of war that is peace.[19]

From his entry into Gethsemani in 1941 to his ordination as a priest in 1949, there is a gap in our knowledge about Merton's involvement in Asian religions. It is understandable that the drastic change from an exceedingly worldly existence to that of a Trappist monk, coupled with the exacting studies in preparation for the priesthood, absorbed all his energies. Yet the monastic discipline, especially the practice of silence, asceticism, and contemplation paved the way for an exploration in greater depth of Oriental monasticism and mysticism. On 4 June 1949, we find the following entry in Merton's monastic journal, *The Sign of Jonas,* on the occasion of the visit of Archbishop Paul Yu-Pin of Nanking to Gethsemani:

> . . . he spoke in Chapter about China and the contemplative life and Buddhist monasticism–and about the reproach that Buddhists fling at us, that is, we are all very fine at building hospitals but we have no contemplatives. He spoke of the two million (or was it five million?) Buddhist monks and nuns. He told of whole mountains covered with monasteries, and then spoke of the immense influence

> exercised by a Christian contemplative community like our Lady of Consolation, when it existed . . . . [20]

On 24 November 1949, the journal referred to the fact that Merton was engaged in the study of Patanjali's Yoga and that someone who had to do a paint job in the monastery had been previously a postulant in a Zen Buddhist monastery in Hawaii.[21] In 1951, in his book, *The Ascent to Truth* there is further evidence that Merton had occupied his mind with Yoga and Zen. He alluded to similarities between the views of St Gregory of Nyssa and St John of the Cross and those found in Patanjali's Yoga and in Zen Buddhism.[22] A formal dialogue on religious experiences in East and West was ushered in in the spring of 1959, when Merton sent to Dr T. D. Suzuki, the chief exponent of Zen in the West, a copy of his manuscript *The Wisdom of the Desert.* A correspondence ensued which will be discussed in the chapter on Merton and Zen.

What prompted Merton to send his manuscript on the Desert Fathers to a Zen master? He had concluded: '. . . these Desert Fathers had much in common with Indian Yogis and with Zen Buddhist monks of China and Japan'.[23] What had attracted Merton to the Desert Fathers and how he viewed their mission throws light on what he expected to obtain from a study of Asian mystics. He emphasized that those who had abandoned the centers of civilization for the desert in the third and fifth century sought 'a society where all men were truly equal, where the only authority under God was the charismatic authority of wisdom, experience and love'.[24] And they had sought 'A life of solitude and labour, poverty and fasting, charity and prayer which enabled the old superficial self to be purged away and permitted the gradual emergence of the true, secret self in which the Believer and Christ were "one Spirit".'[25] By overcoming their alienation from their true self, they gained true freedom to love their fellowmen unconditionally. Thus the flight into the desert was not an escape from the world.

As a matter of fact Merton's attitude towards what constitutes the world had undergone a change in 1958:

> In Louisville, at the corner of Fourth and Walnut, in the center of the shopping district, I was suddenly overwhelmed with the realization that I loved all those people, that they were mine and I theirs, that we could not be alien to one another . . . . It was like waking from a dream of separateness, of spurious self-isolation in a special world, the world of renunciation and supposed holiness . . . . [26]

He was beginning to realize that by living in a monastery one did not cease to be part of the world. In fact, a monasticism that had become conventionalized, routinized and overly structured, partook of many features he had associated with the sinful world. It was now necessary to undertake 'a leap into the void'.[27] This the Desert Fathers had done, as also had the mystics of the East.

Impelled to leap into the void, Merton intensified his efforts at probing the wisdom of Asia. His search for ultimate reality was characterized by a contradiction. On the one hand, he knew that he had to undertake an inner journey; ' . . . that is far more crucial and infinitely more important than any journey to the moon.'[28] He was also aware that the Desert Fathers had acquired their spiritual insights by delving within rather than by intellectual studies and distant journeys. Merton vacillated between an inner and an outer journey, eventually embarking upon the trip to Asia from which he was not to return. From 1961 on he studied Asian spirituality in a systematic manner. Asia became his major concern to the end of his life, as is apparent from some of his significant publications: *The Way of Chuang Tzu* (1965), *Mystics and Zen Masters* (1967), *Zen and the Birds of Appetite* (1968) and finally, the posthumously published *The Asian Journal of Thomas Merton* (1973).

When Merton decided to direct his major energies towards a deeper understanding of Asian religions and philosophies he thought it essential to obtain the guidance of Asian scholars. With the permission of his abbot, Merton wrote to Archbishop Yu-Pin, now at Taipei, Taiwan, who had so greatly impressed him during his visit to Gethsemani in 1949. In his letter to the archbishop, 3 February 1961, Merton expressed his belief that it was God's will that he make the study of Oriental thought his primary occupation, and he asked the archbishop to suggest names of Asian experts who could aid him in his venture.[29] Archbishop Yu-Pin was unavailable at that time, but Merton received a reply from Father Paul Chan, Secretary-General of Sino-American Amity, Inc. in New York. Chan recommended Lin Yutang, Dr Richard S. Y. Chi, Professor of East Asian Languages and Literature at Indiana University, and Dr John C. Wu, Professor of Oriental Philosophy and Religion at Seton Hall University, South Orange, New Jersey.[30]

Professor Wu was especially suited for collaboration with Merton. A jurist, diplomat, scholar and, at the same time, a convert to Catholicism, he became Merton's friend and advisor. From the start Wu expressed his admiration for Merton's insights into Asian thought, which he praised with a touch of Asian hyperbole: 'I thought you were a wizard, Father! You have such penetrating insight into the Chinese ways of thinking . . . you are another Christ. Let me serve you as an altar boy.'[31] There was undoubtedly soul affinity between Merton and Wu. Both were eager to bring Christianity and Asian wisdom closer together. In his book, *Chinese Humanism and Christian Spirituality,* Wu had pointed out similarities in Mencius' philosophy of human nature and the natural law tradition of St Thomas Aquinas. Wu also detected resemblances in the 'little way' of St Thérèse of Lisieux with Confucius' concept of filial piety and with Lao Tzu's insistence on regaining the spirit of childhood.[32] In his introduction to another work of Professor Wu, *The Golden Age of Zen,* Merton wrote: 'Dr. Wu is not afraid to admit that he brought Zen, Taoism and Confucianism with him into Christianity. In fact in his well-known Chinese translation of the New Testament he opens the Gospel of St. John with the word, "In the beginning was the Tao".'[33] In a letter he wrote to Merton on Good Friday 1961, Wu exclaimed: ' . . . *The Tao Incarnate* is absolutely the *Same Tao* who was from the beginning with God and is God. Before Lao Tzu was, He is.'[34] And, Wu added:

> The way to the re-Christianization of this post-Christian West lies through the East. Not that the East has anything really new to give to the Gospel of Christ; but its natural wisdoms are meant to remind Christians of their infinitely richer heritage, which, unfortunately, they are not aware of.[35]

Merton enthusiastically exclaimed on 28 December 1965: 'This is the only real contribution I have to make to a tormented political situation. Instead of fighting Asia to be it. And stubbornly, too.'[36]

Convinced that Asian wisdom could enrich western education and renew Christianity, Merton pointed out that in the formative period of Christianity Greek philosophy had contributed to Christian culture. Could Asian thought not now help bring about the rejuvenation of Christianity, especially since Asia had specialized in contemplation and to this day holds it in high regard? And contemplation he thought essential as a balancing force to western hyperactivity. It was a necessary prerequisite for right action, for purity of heart. If practiced widely, Merton contended, contemplation could overcome racial, social, religious and national strife and thus make possible world peace.[37] But was contemplation really the panacea for all the world's ills? Again, Merton conveniently ignored the fact that a truly contemplative civilization had never existed even in Asia. And while contemplation was undoubtedly highly valued in ancient India, it had not produced a harmonious, peaceful society. On the contrary, at the very time when yoga was widely practiced (ca. fourth century B.C.), an Indian minister of state, Kautilya, composed the *Arthasastra,* a Machiavellian treatise on power politics. It was the *Arthasastra* and not yoga which guided Indian rulers who considered it their sacred duty to conquer neighboring territories. Not by chance has the distinguished Indologist A. L. Basham drawn attention to 'Hindu Militarism'.[38]

Inspired by Wu, his fellow bridge-builder between East and West, Merton undertook a study of Chinese culture. He drew encouragement from his noble predecessors, the Jesuits of the sixteenth century, to whom he paid respects in *Mystics and Zen Masters.* He pointed out that the Jesuits had conceived their task in the Heavenly Kingdom very differently from other missionaries, some of whom had wanted to missionarize China following the example of the *conquistadores* of Mexico and Peru. Considering the Chinese mere pagans, they had advocated the use of armed force. The Jesuits, in contrast, had been appreciative of Chinese culture, which they had studied with the utmost respect. Merton emphasized: 'When Ricci dressed as a Confucian scholar, this was not a Jesuitical disguise. The Jesuits wore the traditional robes of the Chinese scholar because they earned the right to do so just as seriously as any other Chinese scholar.'[39] Matthew Ricci and his associates had realized that Christian missionary efforts could only succeed if Christianity was presented stripped of its western cultural accretions. The Jesuits, following in the footsteps of St Paul, had determined to become 'all things to all men' (1 Co 9:22).[40] It seemed to them only reasonable that the Mass and the breviary be recited in Chinese. Also, they had permitted Chinese converts to continue the observation of family rites and rites in memory of Confucius. In view of the fact that Chinese culture was based on respect for ancestors and on 'filial piety', it was practically unthinkable for a Chinese to abandon customs which, as the Jesuits had realized, were not idolatrous. Merton compared the significance of the 'rites' for the Chinese with American respect for the star spangled banner.[41] Alas, by the second half of the seventeenth century the Jesuits were under attack by rationalists and freemasons as well as from rival missionaries. In particular, they were accused of 'unorthodox behavior', and in 1704 the Chinese rites were banned. The ban was not lifted until 1939. By that time the opportunity for a large-scale conversion of the Chinese to Christianity had been lost. Merton commented that the Jesuit missionaries 'had been three hundred years ahead of their time, were profoundly concerned with issues which we now see to be so important that the whole history of the Church and of western civilization seems to be implicated in their solution.'[41b] He warned:

> If the West continues to underestimate and neglect the spiritual heritage of the East, it may hasten the tragedy that threatens man and his civilizations. If the West can recognize that contact with Eastern thought can renew our appreciation for our own cultural heritage, a product of the fusion of the Judeo-Christian religion with Greco-Roman culture, then it will be easier to defend that heritage, not only in Asia but in the West as well.[42]

Merton went so far as to argue that it was incumbent upon the Christian West to guard the cultural and religious treasures of the East from western materialism.[43]

In his analysis of Chinese classical society Merton concentrated on the two complementary traditions of China: Confucianism and Taoism. He agreed with the Jesuits of the sixteenth century that Confucianism should be viewed as a 'sacred philosophy' compatible with Christianity. In fact, Confucian ethics seemed to him similar to the wisdom literature of the Bible.[44] He was fascinated by Confucius' emphasis on the use of the solemn liturgical rites and music as a way in accordance with the will of heaven of bringing order and harmony to society:

> One might almost say that for Kung [Confucius] 'rites' or *Li* were the visible expression of the hidden reality of the universe: the manifestation of heaven, or, we would say, of divine wisdom, in human affairs and in the social order. It is

> not enough for the divine order to be present metaphysically: or enough for man to bring it into his own life of society by moral conduct. The will of heaven is something that has to be *celebrated* with beauty and solemnity . . . .[45]

Julia Ching in *Confucianism and Christianity: a Comparative Study* has concluded, in this connection, that early Confucianism exhibits similarities with Christianity: ' . . . in its life of prayer and cult as well as in the theology this implies'.[46] In fact, she argued, Confuciansim was more compatible with Christianity than Buddhism.[47]

Confucianism also had an answer for the problem of alienation, Merton contended. Through the faithful observance of rites man would be put in touch with his true self which, in turn, was in tune with the will of heaven. The Confucian ideal of *Jen* (love or human-heartedness) contributed to human harmony. Confucian education, based on *Li* and *Jen,* aimed at achieving a meaningful performance of all actions, no matter how trivial they might be. Merton sang the praises of the Confucian hierarchical social order: ' . . . father to son, marked by *justice,* mother to son, marked by *compassion,* or merciful love; the son to his parents, marked by *filial love;* the elder brother to his younger brother, marked by *friendship;* and the younger to the elder, marked by *respect* for his senior.'[48] He considered the social organization of China 'a wonderful organic complex of strength',[49] even though he had to admit that the lofty concept of filial piety and the Confucian humanistic ideal were frequently ignored by the Chinese emperors and highly-placed officials. That in reality China had all too often been governed by legalism, a Machiavellian system unconcerned with morality, focusing instead on ruthless pursuit of power,[50] did not reduce his enthusiasm for China.

Even more congenial than Confucianism to Merton's way of thinking and feeling was Taoism. In contrast to Confucius, Lao Tzu, the mythical founder of Taoism, rejected a systematic structuring of society and the formation of character through formal education and adherence to precise rites. He contended that the ideal society, based on the goodness of its individual members, could only come about if spontaneity prevailed. To follow one's intuition, without self-conscious reflection, would lead one to harmony. Actually, all that one needed was to be in tune with the *Tao* (Way). But what was the *Tao*? Confucius had also referred to the *Tao,* but, according to Lao Tzu, the Confucian *Tao* was the *Tao* 'that could be named', while his *Tao,* the eternal *Tao,* was nameless. It was indefinable, transcending all concepts, all modes of visualization. At the very moment we attempt to define the *Tao,* we lose it. Similarly, inasmuch as the *Tao* embodied perfection, we lose our perfection when we make a self-conscious effort to practice perfection. Instead we should practice *Wu Wei,* which Merton, following his mentor Wu, translated as 'non-ado'.[51] Merton elaborated: 'Hence *wu wei* is far from being inactive. It is supreme activity, because it acts at rest, acts without effort. Its effortlessness is not a matter of inertia, but of harmony with the hidden power that drives the planets and the cosmos.'[52] It is our task to remove the obstacle to perfect and loving action of the *Tao:* our egoistic cravings which produce 'Much Ado about Nothing'. Then we become perfect instruments for the *Tao* to act through us. By losing ourselves, we find ourselves, i.e., our true identity. The *Tao* proves to be the way to overcome alienation and gain the peace 'that passeth understanding', Merton's major objective. He was impressed by the similarity between Taoist teachings and those of Christ. The sage who has 'found' the *Tao* has detached himself from all material desires; 'The foxes have holes and the birds of the air have nests, but the Son of Man has nowhere to lay his head' (Mt 8:20). Of all the sayings in the *Tao Te Ching* Merton considered the sixty-seventh chapter closest to Christianity: 'Because I am merciful, therefore I can be brave . . . / For heaven will come to the rescue of the merciful and protect him with its mercy.'[53] A passage from the

Tao Te Ching which he recommended for serious consideration to the leaders of the United States referred to the hazards of war:

> To rejoice over victory is to rejoice over the slaughter of men!
> Hence a man who rejoices over the slaughter of men cannot expect to thrive in the world of men.
> . . . Every victory is a funeral.[54]

We must remember that Merton's comments were written at the height of America's involvement in the Vietnamese war, at a time when Merton himself was deeply engaged in the resistance movement to the war.

Even more than to Lao Tzu, Merton was attracted to the other great representative of Taoism, Chuang Tzu (ca. 369–286 B.C.). 'I simply like Chuang Tzu because he is what he is . . . . '[55] Merton further justified his liking for Chuang Tzu:

> If St. Augustine could read Plotinus, if St. Thomas could read Aristotle and Averroes (both of them certainly a long way further from Christianity than Chuang Tzu ever was!), and if Teilhard de Chardin could make copious use of Marx and Engels in his synthesis, I think I may be pardoned for consorting with a Chinese recluse who shares the climate and peace of my own kind of solitude, and who is my own kind of person.[56]

Merton completed his work on *The Way to Chuang Tzu* in 1965, after five years of study and meditation. Writing in close collaboration with Wu, his 'chief abettor and accomplice,'[57] Merton selected for interpretation passages which especially appealed to him. Since he only knew a few Chinese characters—lack of time had prevented him from fulfilling his aspiration of mastering the Chinese language—he had to rely on translations of the sayings of Chuang Tzu into English, French, and German. In his preface to *The Way to Chuang Tzu* Merton indicated that he had enjoyed writing the book 'more than any other I can remember'.[58] Brother Patrick Hart, Merton's secretary, explained that Merton's destription of Chuang Tzu as 'a thinker who is subtle, funny, provocative, and not easy to get at'[59] may be considered 'a mirror image of Merton who doubtless unconsciously identified with him . . . . '[60]

The central concept of Chuang Tzu is the 'complementarity of opposites',[61] the realization that under certain circumstances good can turn into evil, big may be viewed as small, and life as death; that, consequently, one must avoid becoming attached to absolutes which contradict a universe in flux, a life which, at least outwardly, is full of contradictions: 'When a limited and conditioned view of "good" is erected to the level of an absolute, it immediately becomes an evil, because it excludes certain complementary elements which are required if it is to be fully good.'[62] Undoubtedly the concept of the complementarity of opposites appealed to Merton and helped him in coping with his inner contradictions and the contradictions he encountered in the world around him. It prompted his assessment in 'First and Last Thoughts', a preface to Thomas P. McDonnell's *A Thomas Merton Reader:* 'I have had to accept the fact that my life is almost totally paradoxical . . . . Paradoxically I have found peace because I have always been dissatisfied . . . . '[63] He gives at least intellectual assent to the need for accepting a duality which does imply imperfection, as it is seen in Chuang Tzu's saying 'Great and Small':

> Consequently: he who wants to have right
>   without wrong,
> Order without disorder,
> Does not understand the principles
> Of heaven and earth.
> He does not know how
> Things hang together.
> Can a man cling only to heaven
> And know nothing of earth?
> They are correlative: to know one
> is to know the other.
> To refuse one
> Is to refuse both . . . . [64]

In *The Way of Chuang Tzu* Merton continued to look for similarities between Taoism and Christianity. In this connection he referred to St Augustine's statement, 'That which is called the Christian religion existed among the ancients and never did not exist from the beginning of the human race until Christ came in the flesh'.[65] Chuang Tzu assuredly had castigated a society bent on chasing after mammon: 'The rich make life intolerable, driving themselves in order to get more and more money which they cannot really use. In so doing they are alienated from themselves, and exhaust themselves in their own service as though they were slaves of others.'[66] This comes close to Jesus' saying: 'You cannot be slave of both God and of money' (Mt 6:24). But he failed to draw the obvious conclusion that Chuang Tzu's criticism of his contemporary society proved that ancient China manifested similar shortcomings as the modern West. In fact, not only Chuang Tzu, but also Lao Tzu and Confucius, had protested the evils of their time. Had Merton admitted that ancient Chinese society and, for that matter, all societies past and present, exhibited those features which he condemned in his own western society, he would have doomed the utopian expectations which sustained the momentum and enthusiasm of his quest. To protect his cherished hopes, he had to overlook unpleasant reality.

Merton also pointed out the striking resemblance in Chuang Tzu's view of ethical principles with St Paul's attitude towards Old Testament Law. In 'Cracking the Safe' Chuang Tzu argued:

> Moral: the more you pile up ethical principles
> And duties and obligations
> To bring everyone in line
> The more you gather loot
> For a thief like Khang.
> For ethical argument
> And moral principle

> The greatest crimes are eventually shown
> A signal benefit
> To mankind.[67]

St Paul states in Rm 7:7-8: ' . . . I should not for instance have known what it means to covet if the law had not said *You shall not covet.* But it was this commandment that sin took advantage of to produce all kinds of covetousness in me, for when there is no Law, sin is dead.' By concentrating on becoming virtuous we are likely to miss being truly virtuous. Rather should we forget ourselves (our false selves) and act intuitively with spontaneity. Virtue will result from our action, but we shall be unaware of it, since it will be, not an object apart from us, but our very nature: the *Tao.* Rational analysis will not enable us to attain virtue, nor, for that matter, wisdom. In general, Chuang Tzu scoffed at the idea that we can attain any worthwhile goal through a method. In the manner of Jesus Chuang Tzu taught by paradox. Challenging the concept of utility, he proclaimed ' . . . the absolute necessity of what has "no use".'[68] Merton as well as Chuang Tzu must have had in mind the hermit, the contemplative, whom conventional society, practical down-to-earth 'realists' label useless. Similarly, Jesus reprimands the 'realists', citing Isaiah's prophecy:

> You will listen and listen again, but not understand,
> see and see again, but not perceive.
> For the heart of this nation has grown coarse,
> their ears are dull of hearing, and they have shut their eyes,
> for fear they should see with their eyes,
> hear with their ears,
> understand with their heart,
> and be converted
> and be healed by me (Mt 13:15).

This is not far removed from Chuang Tzu's teaching on the fasting of the heart:

> The goal of fasting is inner unity. This means hearing, but not with the ear; hearing, but with the understanding; hearing with the spirit, with your whole being . . . it demands

> the emptiness of all faculties . . . frees you from limitation and from preoccupation. Fasting of the heart begets unity and freedom . . . So when the faculties are empty, the heart is full of light . . . .[69]

The fasting of the heart not only suggests Christ's teachings on poverty of spirit but also resembles the Buddhist concept of *Sunyata* (emptiness). Merton was aware of the close connection between Taoism and Buddhism. He rightly concluded that the spiritual heirs of Chuang Tzu were the Chinese Cha'an (Chinese equivalent for Zen) Buddhists of the T'ang era (the seventh to tenth centuries A.D.). Both Chuang Tzu and Cha'an Buddhists were 'not concerned with words and formulas about reality, but with the direct existential grasp of reality itself . . . .'[70] Direct grasp of reality became Merton's all-consuming passion and led him 'logically' from Taoism to Zen Buddhism.

Chapter II
MERTON AND ZEN BUDDHISM

AMONG ALL ASPECTS of Asian wisdom Zen made the greatest impact on Merton. The night before his fatal accident, in December 1968, Merton told the poet John Moffit: 'Zen and Christianity are the future'.[1] In his book *Zen and the Birds of Appetite* he wrote: 'I believe that Zen has much to say not only to a Christian but also to a modern man. It is nondoctrinal, concrete, direct, existential, and seeks above all to come to grips with life itself, not with ideas about life . . . . '[2] He felt strongly that Zen could contribute to answering four major needs of modern man: 1) the need for harmony among nations and within individual societies, 2) the need for meaning in the ordinary daily activities of life, 3) the need for integrating one's individual, intellectual, and spiritual nature, and 4) the need for finding one's true self as well as for self-transcendence.[3]

Before discussing Merton's views on Zen in detail, we must pause to examine how Merton viewed the relationship of Zen to Buddhism. On the one hand, Merton stated:

> To define Zen in terms of a religious system or structure is in fact to destroy it—or rather miss it completely, for what cannot be 'constructed' cannot be destroyed either . . . . Zen is consciousness unstructured by particular form or particular system, a transcultural, transreligious, transformed consciousness.[4]

On the other hand, he admitted:

> Yet this does not mean that there cannot be 'Zen Buddhists,' but these surely will realize (precisely because they are Zen-men) the difference between their Buddhism and their Zen—even while admitting that for them their Zen is in fact the purest expression of Buddhism.[5]

Does this not sound somewhat like semantic quibbling? The issue is further compounded by his argument that Buddhism is less of an -ism than other religions and that it too aims at liberating men from all structures, forms, and beliefs, so that they can experience undiluted reality. Where then is the difference between Buddhism and Zen? Is there in reality such a thing as 'pure Zen' distinct from Buddhism? Merton conceded: 'The whole study of Zen can bristle with questions like these, and when the well-meaning inquirer receives answers to his questions, then hundreds of other questions arise to take the place of the two or three that have been "answered".'[6] One of the greatest experts on Zen in the Western world, Father Heinrich Dumoulin, considered it problematic to dissociate Buddhism from Zen.[7] Pointing out that Zen derives from Buddhism, he even goes so far as to claim that it is totally Buddhist.[8] While it has been argued that Zen distinguishes itself from Buddhism by its doctrine of 'sudden', rather than 'gradual', enlightenment, Father Dumoulin showed that Indian Mahayana Buddhism conceived of instantaneous enlightenment.[9] It must be noted that in practice Merton himself did not consistently separate Buddhism from Zen and even used the terms interchangeably.[10]

That Merton emphasized the trans-religious nature of Zen was partially due to his wish to allay qualms of conscience on the part of Catholics whom he wanted to reassure that the practice of Zen need not interfere with their religious loyalty. In arguing that Zen was 'non-religious' he was supported by Dr D. T. Suzuki (1870–1966). Suzuki whom Dumoulin considered 'not really a religious type',[11] had been involved in a correspondence with Merton since the later 1950s. Merton had wanted to include an essay by Suzuki on Zen in *The Wisdom of the Desert,* but the monastic censors had excluded it.[12] Suzuki's significance for Merton is indicated by his judgement: 'I can venture to say that in Dr. Suzuki Buddhism finally became for me completely comprehensible, . . . '[13] But Suzuki popularized Zen in the United States while largely disregarding other Buddhist sects. Through Suzuki Merton concentrated on Rinzai Zen and almost totally ignored Zen in its Soto form. While both Rinzai and Soto Zen Buddhists make use of *zazen* (sitting in an upright position with legs crossed in meditation), and pose *koans* (logically insoluble problems), the Rinzai sect emphasizes *koans* and the Soto sect *zazen. Zazen* is generally considered a more gradual procedure whereas the use of *koans* is a more drastic technique aimed at transcending the intellect and evoking pure intuition. The illogical, paradoxical nature of the *koan* aims at jolting the Zen disciple out of his conventional mode of thinking and 'catapulting' him into direct confrontation with reality.[14]

Greatly attracted to Suzuki, Merton naturally craved a personal meeting with the illustrious Zen master, but obtaining leave from Gethsemani was fraught with difficulties. Although Merton had been permitted to go to Louisville in connection with his frequent illnesses, his requests to travel for other purposes had been mostly refused. When, in 1963,

Dom Aelred Graham, the author of *Zen Catholicism* (New York, 1963), had invited him for a visit to the Benedictine Abbey at Portsmouth, Rhode Island, Merton commented: 'I was tempted to reply that it would be easier for me to get permission to take a mistress than it would be to visit a Benedictine monastery. . . . '[15] Finally, in December 1964, Merton was given an opportunity to meet Dr Suzuki at Butler Hall, on the campus of Columbia University. It was a tremendous experience for Merton who placed so much value on experiential learning: 'One cannot understand Buddhism until one meets it in this existential manner in a person in whom it is alive.'[16] Suzuki's charming secretary, Miss Mihoko Okamura, performed the tea ceremony, which deeply impressed Merton. Suzuki's behavior during the ceremoney was more significant than any intellectual exchange that may have occurred with the ninety-four year old Suzuki, who incidentally used a hearing aid which did not work too well.[17] Merton in retrospect reminisced:

> I drink my tea as reverently and attentively as I can. She [Mihoko] goes into the other room. Suzuki as if waiting for her to go, hastily picks up his cup and drains it. It was at once as if nothing at all had happened and as if the roof had flown off the building. But in reality nothing had happened. A very old deaf Zen man with bushy eyebrows had drunk a cup of tea, as though with complete wakefulness of a child and as though at the time declaring with utter finality: 'this is not important'.[18]

Merton added: 'The functioning of a university is to teach a man how to drink tea, not because anything is important, but because it is usual to drink tea, or for that matter, anything else under the sun. And whatever you do, every act however small can teach you everything, provided you know who it is that is acting.'[19] The ceremony can be viewed as a way of sacramentalizing all life. Ordinary functionings, such as drinking and eating can be used to reveal the underlying mystery, their transcendent aspect, the dimension of eternity. It is not surprising that Merton considered the tea ceremony to have features in common with the Eucharist, that those who participate in the ceremony experienced a spirit of communion.[20] Inasmuch as Zen (and the tea ceremoney as an expression of Zen) is tantamount to a transvaluation of ordinary human activities, it can be summed up in the words of the Zen practitioner of the eighth century A.D., Ling-chao, who declared: 'My study is neither difficult nor easy. When I am hungry I eat. When I am tired I rest.'[21] This viewing of the ordinary things in life in an extraordinary way reminds one of Blake's poetic insight:

> To see a world in a grain of sand
> and Heaven in a wild flower,
> Hold infinity in the palm of your hand
> And Eternity in an hour.[22]

But Merton was fully aware that no matter what symbolism one might choose it could not fully express Zen reality. Since Zen was considered indefinable and could not be communicated by words, any discussion of Zen was highly problematical, if not actually absurd:

> The language used by Zen is therefore in some sense an antilanguage, and the 'logic' of Zen is a radical reversal of philosophical logic. The human dilemma of communication is that we cannot communicate ordinarily without words and signs, but even ordinary experience tends to be falsified by our habits of verbalization and rationalization.[23]

The problem of communicating Zen experiences is not greatly different from conveying mystical experiences; in the ultimate sense they are ineffable. Yet mystics have sought to express their experiences through symbols and also often in a language which seems to contradict reason and logic. With all its emphasis on ineffability, Zen had produced thousands of texts expounding the indefinable. In the case of Merton, the problem of verbalization was aggravated by his unquestionably undue verbosity. Sister Elena Malits csc, in her study *The Solitary Explorer: Thomas Merton's Transforming*

Journey, concluded:

> There is a hint of the ludicrous in Merton's constant writing and talking about silence . . . . This Trappist monk produced volumes of print and file cases full of tapes denouncing useless words, unnecessary chatter and continuous verbalizing. Yet he himself could speak quite garrulously and write compulsively.[24]

Dr. Gregory Zilboorg, a Catholic psychiatrist whom Merton met at a workshop on pastoral care and psychiatry in Collegeville, Minnesota, had suggested that Merton become less 'verbalogical'; Sister Malits conjectures that this advice might have contributed towards steering Merton in the direction of Zen.[25] There is no way of determining whether without Zen Merton would have been still more 'verbalogical'. But his life is a reflection of the difficulty facing an intellectual who tries to transcend concepts in order to attain an undiluted experience of reality. He was acutely aware of the temptation to substitute words for reality; he constantly warned his contemporaries against this danger and yet often succumbed to it.

'Undiluted reality' was the goal Merton sought and believed Zen could induce. *Nirvana,* Enlightenment, Purity of Heart, 'Life in the Spirit', Buddha Mind, *Sunyata,* these were the various terms Merton used often interchangeably to describe the indescribable. He was convinced that when one experienced 'undiluted reality', one was freed from all delusions, including one's ego consciousness. One was in harmony with everyone and everything, and one consequently experienced life 'as it really is', thus fulfilling modern man's need for harmony, self-transcendence and ultimate meaning. On the other hand, Zen masters warned that in reality there was no method which enabled one to attain 'undiluted reality', no one to experience it, and nothing to be experienced. Yet Zen practitioners underwent austere disciplines to rid themselves of all conventional obstacles, of common human misconceptions which had caused deep grooves in their consciousness, in order to break through into ultimate reality.[26]

To illustrate Zen reality, Merton—in agreement with Father Dumoulin—pointed to the differences between the Northern school of early Chinese Zen, represented by Shen-Hsiu (606-706 A.D.) and the Southern school represented by Hui Neng (638-718 A.D.), one of the outstanding exponents of Zen. It is reported that the leader of the Zen school, the Fifth Patriarch Hung-jen (610-674 A.D.) ordered his disciples to compose a *gatha* (stanza). The disciple whose *gatha* exhibited the greatest degree of enlightenment was to become his successor. Shen Hsiu wrote:

> The body is the Bodhi tree [enlightenment]
> The mind is like a clear mirror standing
> Take care to wipe it all the time
> Allow no grain of dust to cling.[27]

When Hui Neng, who incidentally was illiterate, heard about Shen Hsiu's stanza, he composed the following response:

> The Bodhi is not like a tree
> The clear mirror is nowhere standing
> Fundamentally not one thing exists;
> Where, then, is a grain of dust to cling?[28]

Hui-Neng became the Sixth Patriarch and caused thereby such resentment among the sympathizers with Shen Hsiu that Zen split into a Northern and Southern school. As one might have expected, Merton preferred Hui Neng to Shen Hsiu. Commenting on the two stanzas Merton observed:

> . . . for Shen Hsiu, the enlightenment and 'seeing' of Zen consists in an awareness of primal mirror-like purity, and the 'mirror light' of the mind is the basis or 'stand,' upon which contemplation solidly rests . . . . It is a primordial reality to be sought as an objective basis for contemplation. For Hui Neng there is no primal 'object' on which to stand, there is no stand, the 'seeing' of Zen is a non-seeing . . . .[29]

He concluded that Hui Neng's approach was analogous to the Desert Fathers' apophatic way (from the Greek *apophasis* meaning denial).[30]

Comparing Zen with the approach of the Desert Fathers, Merton realized the necessity of elucidating the relationship between Christianity and Zen Buddhism. In his excellent essay, "The Zen Catholicism of Thomas Merton,' Professor Chalmers MacCormick analyzed the gradual change in Merton as he moved from considering Zen distinctly inferior to Christianity to a greater appreciation of Zen, and ultimately to an admission that at least individual Zen masters might have had supernatural experiences equivalent to those of Christian mystics.[31] Within the theological dimension, Merton invariably stressed the marked doctrinal differences between Christianity and Buddhism. He pointed out that 'in Christianity the objective doctrine retains priority both in time and in eminence. In Zen the experience is always prior, not in time but in importance.'[32] Buddhism and Biblical Christianity agreed in their conclusion that mankind's suffering resulted from a distorted view of reality, he averred. The Buddhists explained man's condition in terms of *avidya* (ignorance). In our ignorance we look at all of life's happenings from an egocentric view, reacting with pleasure and displeasure, depending whether the ego interprets occurrences positively or negatively. Since the phenomenal world is in constant flux, egocentricity prevents us from attaining lasting peace, for we alternately experience feelings of attraction and repulsion, joy and sorrow in this world of duality. Christians explain mankind's dilemma in terms of original sin and man's fall. Mankind selfishly tries to manipulate the world in defiance of God and divine laws. There is a significant difference between original sin and *avidya* which Merton failed to emphasize. The suffering resulting from the fall of man has been brought about by man himself, while no responsibility for *avidya,* the cause of suffering in Buddhism, is laid upon man.[33]

The Buddhists try to cope with man's condition by advocating right mindfulness; through meditation man becomes aware that he does not possess a separate self and he eventually attains the state of *nirvana,* which Merton defined as perfect awareness and perfect compassion: ' . . . the wide openness of Being itself, the realization that Pure Being is Infinite Giving, or that Absolute Emptiness is Absolute Compassion'.[34] Given the distinction between original sin and *avidya,* one might ask, can Buddhist meditation really cope with original sin? Merton argued that Zen meditation shatters the false self and restores us to our paradisical innocence which preceded the fall of man.[35] Here again he expressed a utopian view, for there is no evidence that a society living in paradisical innocence has ever existed. In no way did he consider *nirvana* negatively, a denial of life. He knew that the Buddhists viewed the attainment of *nirvana* as a transformation of consciousness excluding the Christian concept of grace. Yet he argued, ' . . . one can hardly help feeling that the illumination of the genuine Zen experience seems to open out into an unconscious demand for grace—a demand that is perhaps answered without being understood. Is it perhaps already grace?'[36]

The Christian answer to the human condition is to be found through the way of the cross leading to ' . . . death and resurrection in Christ—a life of love "in the spirit".'[37] The way of the cross has nothing rational or logical about it: 'The language of the cross may be illogical to those who are not on the way to salvation, . . . ' (1 Co 1:18). And the way of the cross too demanded the death of the ego. At the same time it seemed to Merton that Christianity went beyond Buddhism by positing the ultimate goal of the Kingdom of God. In a dialogue with Dr Suzuki he explained: ' . . . the work of the *new* creation, the resurrection from the dead, the restoration of all things in Christ. This is the real dimension of Christianity, the eschatological dimension which is peculiar to it, and which has no parallel in Buddhism.'[38] But in response Dr Suzuki reminded him that Christian theology included the concept of 'realized eschatology', that 'Paradise has never been lost and therefore is never regained'.[39] To this Merton assented. He eventually concluded that the theological comparison of Buddhism and Christianity was a blind alley: ' . . . to what extent does the theology of a theologian without experience claim to interpret correctly the "experienced theology" of the mystic . . . . '[40] Merton knew that this question cannot be satisfactorily answered. Therefore he proposed experience as the basis of a comparison of religions. After all, not only Zen but also Christianity, although it stressed doctrine, was based on experience: ' . . . people forget that the heart of Catholicism, too, is a *living experience* . . . . '[41] It is founded on the life of Christ as presented in his teachings, and it enjoins his disciples to become Christ-like: 'You must therefore be perfect just as your heavenly Father is perfect' (Mt 5:48); or, as St Paul put it: 'Let your armor be the Lord Jesus Christ; forget about satisfying your bodies with all their cravings' (Rm 13:14). Similarly, the Buddhist is admonished to attain the mind of Buddha. William Johnston in *The Inner Eye of Love: Mysticism and Religion* points out: 'When Jesus and the Buddha meet in their disciples, real mystical dialogue will have begun'.[42] Johnston rightly distinguishes between belief and naked faith. While he considers belief something external, clothed in words and culturally conditioned, he regards naked faith as beyond 'thoughts and images and concepts'.[43] It is naked faith, he avers, which is the essence of all religions. Beliefs are merely their outward restricted expression.[44] From this follows the limitation of comparative dogmatic theology which deals with the culturally determined in contrast to spiritual theology which transcends culture.[45] The consequences of an experiential comparison of religions is aptly summarized by Daniel J. Adams in his article 'Methodology and the Search for a New Spirituality':

> In this way a common ground can be reached which will provide a foundation for later doctrinal discussions. Focus on what is universally human—the content—and then try to understand that content within the framework of the

> cultural form. The result will be a new appreciation of spiritual—and human—experience.[46]

Seeking similarities between Christian and Buddhist religious experiences led Merton inevitably to the Rhineland mystic, Meister Eckhart (1260–1326). Merton's growing appreciation of Meister Eckhart was partially due to Dr Suzuki's praise of Eckhart in his essay, "Meister Eckhart and Buddhism'.[47] Not by chance had Suzuki selected Eckhart as a representative of Christian mysticism when he wanted to compare Christianity with Buddhism. Rudolf Otto before him had chosen Eckhart when discussing similarities in eastern and western mysticism. Ananda Coomaraswamy, who influenced Merton's understanding of Hinduism, saw an analogy between the thought and experience of Eckhart and that of Hindu sages. Eckhart was not an altogether typical representative of Christianity; some of his views were declared heretical and pantheistic. On the other hand, many a mystic has been suspected of pantheistic tendencies, not so much because they actually were pantheists but because they expressed their mystical experiences in terms that suggested pantheism. In the case of Eckhart, Merton emphasized, one must keep in mind that his condemnation was at least in part due to the rivalries between Franciscans and Dominicans, and that the teachings of the Dominican Eckhart were based to a great extent on those of St Thomas Aquinas.[48]

In comparing Eckhart's experiences with those of Zen masters Merton admitted that there existed a basic methodological problem. How could one compare experiences? Obviously, one could only compare the verbal expression of experiences. And the verbal expression of experiences is conditioned by the language and the theological presuppositions of the experiencer. In this connection Merton warned against the common glib assumption that mystics have identical experiences and that 'All religions thus "meet at the top".'[49] All comparisons have therefore to be undertaken with great caution, but, for the sake of furthering mutual understanding between the various religions, the venture had to be carried out. Sister Thérèse Lentfoehr SDS, a close friend of Merton, aptly explained what especially attracted him to Eckhart: 'In Eckhart, Merton found a like attempt to delineate the experience of this inner awareness of God, and he saw parallels between the metaphysical intuition of Zen—its fullness, limitlessness, and utter freedom—and the experience of the mystics'.[50] Sister Thérèse also draws attention to the fact that some of Merton's zen-like poetry was influenced by Eckhart, citing as an example, a poem based on Eckhart's sermon 28, 'Blessed are the Poor':

> When in the soul of the serene disciple
> With no more Fathers to imitate
> Poverty is a success,
> It is a small thing to say the roof is gone:
> He has not even a house.[51]

Analyzing Eckhart's sermon 28, and thereby also explaining the underlying meaning of his own poem, Merton focused on a key passage in the sermon: ' . . . a man should be so poor that he is not and has not a place for God to act in. To reserve a place would be to maintain distinctions'.[52] He interpreted utmost poverty (*eigentlichste Armut*) to signify nothingness, i.e., the elimination of the ego. That state, equivalent to *Sunyata,* enabled man to be filled with God, the Absolute Ground of Being. Merton warned that as long as we imagine that it is we who empty ourselves and make room for God, we are still in ego delusion. All concepts of God and self must vanish. Only pure isness, suchness, or Pure Being must remain. Thus freedom is obtained, for all concepts, all structures, all limitations are gone. The limited ego has made way for limitless Godhead. He concluded:

> . . . when one breaks through the limits of cultural and structural religion—or irreligion—one is liable to end up, by 'birth in the Spirit,' or just by intellectual awakening, in a simple void where all is liberty because all is the actionless action, called by the Chinese *Wu-wei* and by the New Testa-

> ment the 'freedom of the Sons of God'. Not that they are theologically one and the same, but they have at any rate the same kind of limitlessness, the same lack of inhibition, the same psychic fullness of creativity . . . [53]

Further commenting upon Meister Eckhart's concept of the absolute he quoted from Eckhart's sermon 23, 'Distinctions are lost in God: . . . there is something in the soul so closely akin to God that it is already one with him and need never be united with him.'[54] In other words, when there is absolute poverty, the sham ego has died and one's true but so far hidden nature has emerged and manifests the living presence of God within us. All this seemed very zen-like to Merton and similar to 'the discovery not that one *sees* Buddha but that one *is* Buddha', when one has discovered the 'original face before you were born'.[55]

Elaborating on absolute poverty Merton entitled one of his poems *Gelassenheit,* Eckhart's term for letting go (letting God?):

> Desert and void. The Uncreated is waste and emptiness to the creature. Not even sand. Not even stone. Not even darkness and light. A burning wilderness would at least be 'something'. It burns and is wild. But the Uncreated is no something. Waste. Emptiness. Total poverty of the Creator: yet from this poverty springs *everything.* The waste is inexhaustible. Infinite Zero . . . . [56]

As Merton explained elsewhere: 'When the double equation, zero = infinity and infinity = zero, is realized, we have the *eigentlichste Armut,* or the essence of poverty'.[57] This is not really so far removed from Christ's saying: 'Anyone who finds his life will lose it; anyone who loses his life for my sake will find it' (Mt 10:39). Merton drew attention to Christ's self-emptying, his kenosis, as another way of describing nothingness, emptiness.[58] Merton's conclusion is shared by two modern Japanese Buddhists, Maso Abe and Keiji Nishitano.[59] William Johnston also refers to the fact that St Paul's statement: 'Jesus emptied himself' (Ph 2:7) is translated in the

Japanese Bible as: 'Jesus became *mu* (*mu to sareta*): Jesus became nothing'.[60] But, inasmuch as Johnston concedes that Oriental nothingness cannot be fully understood,[61] and one is entitled to add that Christ's self-emptying escapes our understanding, both self-emptying and *Sunyata* remain mysteries.

Merton returned to comparing Zen and Catholic theology in connection with Meister Eckhart's saying: 'In giving us His love God has given us the Holy Ghost so that we can love Him with the love wherewith He loves Himself'.[62] Dr Suzuki had translated this into Zen language as ' . . . one mirror reflecting another with no shadow between them'.[63] Suzuki's translation appeared quite plausible to Merton and prompted him to reaffirm that a fruitful field for future investigation would be comparative spiritual theology rather than speculative theology.[64]

Modern western man's difficulty in understanding Eckhart's as well as the Zen master's concept of emptiness and their ways of communicating spiritual experiences in general Merton blamed chiefly on the Cartesian mentality. He argued that Descartes had reversed reality with his dictum: *Cogito ergo sum*; the ego, the intellect, was considered the measure of all things. By relegating everyone and everything to the status of an object, the ego claimed to be the only subject, thus pitting itself against the whole world. Having converted even God into an object, modern western man took the next logical step and declared the death of God. In reality man ought to proclaim the truth: *Sum ergo cogito,* I am existentially aware of my being, and from that my thoughts ensue.[65] As a result of Descartes' concept of reality, thought had been substituted for reality and subject and object had been split. It seemed to Merton that an injection of Zen into western thinking would be the best way of rectifying the Cartesian error and thereby of healing the western mind. For, as he explained in discussing the Death of God theology, God had only died for the so-called believers, because they had retained a lukewarm, intellectual acceptance of God as a concept which in no way influenced their lives. A marked transformation of their consciousness could lead them to a direct experience of the living God. Such an experience would drastically affect their lives.[66]

Convinced that Zen could contribute to curing western over-intellectualization, Merton reiterated Zen's advantages. He contrasted Zen's opposition to words with Christianity's emphasis on proclaiming the Word of God, the Good News. Zen taught nothing, had no message to convey, except an 'awareness that is potentially already there but is not conscious of itself . . . awareness of the ontological ground of our being here and now . . . ' .[67] Since Zen is *realization,* it is comparable with Christian revelation. In agreement with Aelred Graham,[68] Merton therefore advocated the use of Zen techniques, above all the *koan,* as a means of liberating modern man from his ego-centered, object-oriented thoughts and concepts.

With all his enthusiasm for Zen, Merton was aware that many Westerners misunderstood and misapplied Zen. In particular he was concerned about the Zen wave which had overtaken the United States during the 60s, when many hippies interpreted Zen to mean freedom from moral laws and equated freedom with license, conveniently ignoring Zen's emphasis on discipline. Merton complained: 'Zen has, indeed, become for us a symbol of moral revolt . . . it presupposes . . . freedom from passion, egotism and self-delusion'.[69] Merton was also apprehensive of a mere outward adoption of Buddhist symbolism. When he heard that an American Catholic had immolated himself in protest against the Vietnam War, in imitation of Vietnamese Buddhist monks, he condemned this action as 'inappropriate' for Americans.[70]

While considering certain external imitations of Zen unsuitable for westerners, Merton, to the end of his life, believed that the transformation of personal consciousness through Zen would bring about a more equitable, peaceful society. Yet he acknowledged that Zen itself had stagnated since the Middle Ages and was itself in need of renewal.[71] At the same time he ignored the fact that even within the Asian cultural context the practice of Zen had never produced an equitable, peaceful society but had brought peace and harmony only to individuals. Merton disregarded the fact that Zen in Japan was closely connected with the warrior class (*samurai*). Merton's mentor, Dr Suzuki, had written extensively about Zen and swordsmanship, pointing out that Japanese warriors had been attracted to Zen which had helped them to cope with living under the shadow of death. By means of Zen the warriors had raised themselves above all duality, thus scoffing at the fear of death. In sword fighting they had made use of Zen to overcome ego consciousness and attain a state of spontaneity, to equip them to wield their deadly weapons with lightning speed, without mental hesitation. Moreover, during the Ashikaga era (1338–1573) Zen priests had been active in politics and trade, not necessarily in an edifying manner.[72] Father Dumoulin commented upon the marriage of Zen and the warriors: 'The Japanese knights practised the basic principle of Zen, namely transcendence of life and death. Obviously, fearlessness in the face of death is not the highest human virtue . . . . The play with life and death can degenerate into intolerable cruelty.'[73] Of course, it is not fair to blame Zen for the inhumane warfare during the Japanese Middle Ages, just as the teachings of Christ cannot be held responsible for atrocities committed during the Crusades. If questioned, Merton would probably have interpreted cruelties committed in the name of Zen as aberrations and reaffirmed that the ultimate goal of Zen as well as Christianity was perfect compassion, perfect love, and that Christianity could more easily regain its vitality through Zen's non-speculative method of cleansing one's being. No confrontation with reality could shake his faith in the efficacy of Zen.

Chapter III
MERTON AND HINDUISM

IN 1967, a year before his death, in a posthumously published work, *Opening the Bible,* Merton explained that he preferred Zen to Hinduism because it was 'a completely non-speculative and non-systematic way of direct vision of the ground of being . . . . ' [1] Hinduism he considered too speculative.[2] And yet he had great appreciation for Hinduism, although he never immersed himself in the study of it to the degree he did of Zen Buddhism. He was aware of the fact that Zen itself pointed towards India, for the very word is derived from the Sanskrit term *dhyana* signifying meditation. In *Opening the Bible* Merton praised the Upanishads, ' . . . perhaps the most profound contemplative collection of texts ever written, dealing with the metaphysical unity of

Being and with the yogic consciousness of that unity in concentration and in self-transcending wisdom (samadhi)'.[3] While he was deeply impressed by the lofty, metaphysical contents of the Upanishads, he was primarily influenced by individual exponents of Hinduism.

Merton was first exposed to Hinduism in 1937 while reading Aldous Huxley's *Ends and Means.* While Huxley had not yet become a committed adherent of the Vedanta philosophy, he was already well-acquainted with some of the chief teachings of Hinduism. Through *Ends and Means* Merton became, at least vaguely, familiar with such basic Hindu concepts as non-attachment, delusion, and *atman* (true self, soul), as well as with various branches of yoga.[4] He also understood the use of negation and paradox for expressing ultimate reality. During his studies at Columbia University he had met Dr Mahanambrata Brahmachari,[5] a disciple of Prabhu Jagatbandhu (1871–1920) of Faridpur (East Bengal, now Bangladesh). Prabhu Jagathbandu had started a new *Vaishnava* sect (followers of the deity Vishnu, the preserver), emphasizing the need for celibacy, non-violence, and the removal of caste distinctions. He advocated chanting the divine name of Hari (another name for Vishnu) as a means of liberating mankind from suffering.[6] In 1932 this sect was invited to send a representative to the World Congress of Religions in Chicago. Brahmachari, a twenty-eight year old monk with an M.A. degree from the University of Calcutta, was selected. Arriving too late to attend the congress, he was completely devoid of financial means but managed to make a living lecturing on religion. He even acquired a Ph.D. from the University of Chicago.[7] Through his fellow student, Seymour Freedgood, Merton first met Brahmachari in June, 1938, at Grand Central Station.[8] He was immediately attracted to Brahmachari, sensing that he was in the presence of a God-centered man whose lifestyle starkly contrasted with his own sense indulgence, which he aptly described by saying in retrospect:

> . . . it was this strange business of sitting in a room full of people and drinking without much speech, and letting yourself be deafened by the jazz that throbbed through the whole sea of bodies . . . it was a strange animal travesty of mysticism, sitting in those booming rooms, with the noise pouring through you, and the rhythm jumping and throbbing in the marrow of your bones . . . .[9]

As Merton got to know Brahmachari better, he became more and more impressed by his simplicity, asceticism, and naturalness. Brahmachari made no attempt to convert Merton to Hinduism. Rather he urged him to study the lives of the great Christian saints.[10] In fact, many of Brahmachari's American friends converted to Catholicism. His example helped to mold Merton's attitude toward inter-religious dialogue as well as his eventual outlook on missionary enterprise. Rather than to destroy someone else's value system, he came to feel, one ought to draw out the best from within that other person, who, after all, is our brother. Writing in retrospect in 1964, in tribute to Brahmachari's sixtieth birthday, Merton commented that his encounter with Brahmachari had occurred at a crucial point in his life and had contributed to his eventual decision to become a monk. It was not so much what Brahmachari had said to him, but his very being which had profoundly affected him—the most effective way of teaching. Brahmachari's life of prayer, meditation and worship seemed to Merton infinitely more meaningful than that of the average modern American whose life was dedicated to the frantic pursuit of sense pleasures. Brahmachari had taught him respect and a true love for persons of other cultures: ' . . . to love one's fellow man consists not in depriving him of his own proper truth in order to give him yours, but rather to enable him to understand his own truth better in the light of yours.'[11] Above all he learned that he 'must entrust himself to a higher and unseen wisdom, and that if one can abandon his frantic hold on illusory securities of everyday material existence and entrust himself peacefully to a supreme Will, he will himself find freedom and peace in that Will.'[12]

Around the time of his meeting with Brahmachari Merton became acquainted with the writings of another Hindu, Ananda Kentish Coomaraswamy, whose ideas were greatly to affect Merton's intellectual development, in particular his outlook on art and civilization. Born 22 August 1877, in Colombo (Sri Lanka), the son of a Tamil Hindu and British mother, Coomaraswamy had been barely two years of age when his father died. He was then brought up by his mother in England, where he eventually obtained a Doctor of Science degree from the University of London. From 1906 on he travelled extensively in India, immersing himself in Indian culture. In 1917 he was appointed research fellow in Oriental art by the Boston Museum of Fine Arts, where he remained to the end of his life in 1947, building up and interpreting one of the outstanding collections of Indian Art in the West.[13] In his lectures and in his writings, especially *The Dance of Shiva, Transformation of Nature in Art,* and *Am I my Brother's Keeper?,* Coomaraswamy deplored the materialism of modern western civilization and its destructive influence on traditional eastern society. He pointed to the replacement of indigenous handicrafts by inferior mass produced articles. And he hurled at the West the accusation: 'What you call your "civilized mission" is in our eyes nothing but a form of megalomania.'[14] Yet his condemnation of the West was confined to the modern era, characterized by the divorce of religion from life and art. In the Middle Ages Europe had been essentially spiritual, similar to Asia; 'But subsequent to the extroversion of European consciousness and its preoccupation with surfaces, it has become more and more difficult for European minds to think in terms of unity, and therefore more difficult to understand the Asiatic point of view.'[15] Coomaraswamy had highest praise for the spirituality of ancient India whose 'perennial wisdom' was meant for all mankind. Among representatives of the traditional, spiritually-oriented West, he singled out William Blake and Meister Eckhart, both of whom had a great significance for Merton. In fact, it was in connection with his research for his M.A. thesis on William Blake that Merton, in 1938, first came in contact with the works of Coomaraswamy.

Writing to Coomaraswamy's widow, Dona Luisa, on 13 January 1961, Merton mentioned that while working on Blake, her husband's *Transformation of Nature in Art* 'was decisive in leading me to take the right turn in life and to set foot upon the spiritual road which led to the monastery and to the contemplative life'.[16] Merton regarded Coomaraswamy as a role model, combining the best of East and West, a pioneering crusader for a life of contemplation and simplicity, in opposition to modern 'progress', with its crass materialistic, quantitatively-oriented *Weltanschauung.* Merton's appreciation of non-christian religions, his guilt over western imperialism, and his admiration for pre-modern value systems were undoubtedly greatly reinforced by contact with Coomaraswamy. Merton eagerly accepted Coomaraswamy's idealized view of the Indian social order, according to which the various hierarchically organized castes harmoniously cooperated with each other. Coomaraswamy's contention that each caste 'from that of the priest and the king down to that of the potter and scavenger, is literally a priesthood'[17] is a travesty of reality. But it must have appealed to Merton, especially when he contrasted it with his own assessment of western class society.

Although Merton never carried out his plan of writing a study about Coomaraswamy, he assured Dona Luisa: 'I do hope however that I will always work with something of his spirit'.[18] Merton's wholesale condemnation of modern civilization shared in the spirit of Coomaraswamy:

> . . . the greatest sin of the European-Russian-American complex which we call 'the West' (and this sin has spread its own way to China), is not only greed and cruelty, not only

> moral dishonesty and infidelity to truth, but above all *its unmitigated arrogance towards the rest of the human race.* Western civilization is now in full decline into barbarism . . . because it has been guilty of a twofold disloyalty: to God and to Man.[19]

Direct references to Coomaraswamy's ideas are found in Merton's *Conjectures of a Guilty Bystander,* in *Gandhi and Non-Violence* and in his M.A. thesis, 'Nature and Art in William Blake', (Columbia University, 1939).[20] In this thesis Merton drew attention to the fact that Blake's views on art were influenced by Hindu rather than Greek ideas.[21]

Merton had first been exposed to Blake by his father at the age of ten. At sixteen, while studying at Oakham, England, he developed a strong liking for Blake, although he was far from grasping the depth of Blake's insights.[22] Gradually, Blake's spirituality penetrated his inner being and contributed to his own awakening. He acknowledged that: ' . . . through Blake I would one day come in a roundabout way, to the only true Church . . . '[23] And he exclaimed: ' . . . what a thing it was to live with the genius and holiness of William Blake that year . . . '[24] In this connection with Blake Merton discussed Indian influence on the Hellenistic civilization radiating from centers such as Palmyra and Alexandria. He mentioned that Appollonius of Tyana had studied under a Hindu monk in India's ancient forest academy Taxila. Blake, as he showed, had not only been acquainted with ancient India's cultural impact on the Near East but, through the translations by the renowned Sanskrit scholars William Jones and Charles Wilkins, he had become aware of the spiritual treasures of India. Blake had even read the *Bhagavad Gita,* published in English in 1785. Greatly infatuated with India, Blake claimed that 'Greece stole her art from ancient Hindu monuments'.[25] This is not borne out by modern research, which indicates that although it is probable that Indian philosophy influenced Hellenistic thought, there was hardly any Indian impact on Hellenistic sculpture or architecture. On the contrary, Indian sculpture, especially at Gandhara, was significantly influenced by Hellenistic art forms.[26]

Under the influence of Coomaraswamy Merton concluded that Blake's concept of art was close to that of ancient India. The ideal Indian artist was a practicing yogi who used the power of concentration to attain union between himself and his object. As evidence Merton cited Blake's statement: 'As the eye, such the object'.[27] Merton also detected a similarity between Blakean and Indian concepts of art and that of the scholastic era in the West, drawing attention to Thomas Aquinas' identification of the knower and the known: 'knowledge comes about in so far as the object is within the knower'.[28] To his own satisfaction Merton had substantiated Coomaraswamy's claim that East and West had held similar views on art in the past and that Blake as a mystic in defiance of modernism had shared that view.

Not long after the completion of his M.A. thesis on Blake Merton entered the Trappists. Eight years later, in the fall of 1949, we find the first reference to a resumption of an interest in Hinduism. Merton started a correspondence with a Hindu monk about the *Yoga Aphorisms of Patanjali,* a classical manual on yoga.[29] As he studied Patanjali's yoga, he discerned similarities between Patanjali and St Gregory of Nyssa and St John of the Cross.[30] During the 60s he continued his studies of Patanjali whom he called 'the greatest yogi'.[31] He took extensive notes on the eight steps of Patanjali's yoga leading to the stilling of the mind.[32] Merton emphasized that Patanjali's ultimate goal was seedless absorption (*nirbikalpa samadhi*), i.e., an absorption which transcends all desires and concepts—a state of pure consciousness, in contrast to absorption with seed (*sabikalpa samadhi*), in which traces of desires and concepts still prevent a pure consciousness devoid of subject and object. The ultimate goal of the yogis appeared to him similar to that of the apophatic mystics of the West, for he no longer believed that eastern mystics were not true mystics. He found it reassuring that such distinguished Catholic theologians as Jacques Maritain and Reginald Garrigou-Lagrange, had reached similar conclusions. And he suggested that it was not fair to criticize Hindu yoga on the basis of Aristotelian metaphysics.[33]

In a talk on yoga to contemplative nuns in 1968, Merton contended that yogis had attained a degree of mind control unknown in the West. He maintained that they were able to control their heartbeat as well as their breath. By mentally detaching themselves from their bodies and focusing on their *atman,* they were able to unite with the Divine. He emphasized that yoga was not reserved for an elite but was open to everyone. In this connection he praised the Hindu system of dividing life into four phases: 1) the young celibate student (*brahmacharin*), 2) the householder (*grhastha*), 3) the forest hermit (*vānaprastha*), and 4) the homeless wanderer (*sannyasin*). At the age of seven or eight the young celibate student receives a grounding in the spiritual life under the supervision of a guru. At approximately the age of twenty he enters the householder stage, fulfilling family and civil duties while still practicing yoga. When the continuity of the family has been guaranteed through the birth of grandsons, the householder may leave his home for the forest, intensifying his spiritual practices until in very old age, in anticipation of the end of his earthly existence, he may become a homeless wanderer, completely detached from earthly ties.[34] That in reality many Indians have not followed this ideal scheme, even in ancient times, was overlooked by Merton.[35] He felt that the Catholic church had made insufficient provisions for a semi-monastic life for householders—a kind of Catholic forest hermitage was his ideal. His own observations had convinced him that some married people in their later years were ready for a monastic existence.[36]

While praising the virtues of yoga, Merton warned against irresponsibly mixing Christianity and yoga. He did, however, approve the experiments in Christian yoga by the Benedictine monk, Dom J.-M. Déchanet, who used techniques for breath control, bodily postures and methods of concentration to attain a more authentic Christ-centeredness. As long as the goal was Christ-centeredness, Merton saw no objection to learning from the Hindus.[37] At the same time he cautioned gullible Westerners that not all Indian yogis were truly spiritual and therefore competent guides. Genuine gurus were rare, even in India.[38]

Among yogis who deeply impressed him were Kabir and Ramana Maharshi. Reviewing *One Hundred Poems of Kabir,* as translated by Rabindrananth Tagore, Merton expressed his admiration for the profundity and sincerity of the eclectic Hindu Kabir (1440–1518), who reminded him of the thirteenth-century Dutch mystic, Hadewijch.[39] He was still more effusive in praise of Ramana Maharshi (1879–1950), one of the greatest sages of modern India. In a letter 8 May 1965, to Arthur Osborne, a disciple of Ramana Maharshi and editor of *The Collected Works of Ramana Maharshi* (London, 1959), Merton stated: 'He is a saint to whom I offer the homage of deep admiration . . . . '[40] And he referred to Ramana Maharshi as an authentic modern Desert Father.[41] What especially impressed Merton was Ramana Maharshi's relentless question: 'Who am I?' After eliminating all temporary, superficial attributes of the ego the questioner was led to the true Self, which is equivalent to *Sat-Chit-Ananda* (Existence-Consciousness-Bliss), a Sanskrit term for the Absolute or Ultimate Reality. Through his own contemplative experience Merton had come to the realization that identification with the ego entailed identification with ephemeral rank, position and personal characteristics. But the goal of the contemplative was to advance from the thought 'I am this' to the pure consciousness 'I am'.[42] Merton must have been aware that personal identification with Ultimate Reality constituted a pure monism which Christians ordinarily consider heretical. If only one exists, all distinctions between man and God are obliterated and the existence of a personal God and autonomous individuals is denied. It appears that Merton no longer regarded a monistic view incompatible with the teachings of Christ. His approval of absolute monism may have been encouraged by his acquaintance with the work of the French Benedictine monks Jules Monchanin and Henri Le Saux (the latter better known by his Hindu name Swami Abhishiktananda), who had founded a Hindu-Christian ashram at Shantivanam in Tiruchiapalli, South India, in 1950. Abhishiktananda had concluded, on the basis of his personal experience, that Hindu monism (*advaita*) had its equivalent in Christianity.[43] He drew special attention to Jn 17:21: 'May they all be one, Father, may they be one in us, as you are in me and I am in you . . . .'

Studying the life of Ramana Maharishi helped Merton better understand the role of the guru in Hinduism. He realized that a guru such as Ramana Maharishi taught his disciples by his very being more than through oral arguments. As a matter of fact, Ramana Maharshi observed silence most of the time. Indians yearn to be in the presence of a guru, to walk with him, to meditate close to him and to observe his daily behavior. They assume that the guru can effect subtle, spiritual changes in the disciples and deepen their contemplation. Considering the guru a living embodiment of divinity, the disciples are expected unconditionally to submit to him. Merton compared this demand with the obedience a Cistercian novice owes his abbot or novice master.[44] He emphasized, however, that Christians had no need for a guru 'in the flesh', because they had Christ, a living master. And he considered venerating the Blessed Sacrament equivalent to sitting at the feet of a guru.[45]

A study of guru-disciple relationship is inconceivable without reference to that work which is regarded as the epitome of Hinduism, the *Bhagavad Gita*. In it the divine teacher, Lord Krishna, instructs the ideal disciple, Arjuna. Merton proposed that the *Bhagavad Gita* be made a part of any standard humanities curriculum in the West, side by side with the works of Plato and Homer.[46] He particularly valued the continued vitality of the *Gita,* which had inspired Mahatma Gandhi and the Mahatma's spiritual successor, Vinoba Bhave (1895–), the 'Walking Saint of India'. Merton was thoroughly acquainted with Bhave's *Talks on the Gita;* his copy of it is heavily marked and underlined. Merton contrasted the profound spirituality radiating from the *Gita* and influencing many contemporary Indians with the 'Death of God' philosophy of the materialistic West. In the *Gita* he perceived similarities with the Gospels. Both taught men ' . . . to live in awareness of an inner truth . . . . In obedience to that inner truth we are at last free.'[47] The *Gita,* he contended, demanded a purity of love resembling that preached by St Bernard and by Tauler.[48]

The main theme of the *Gita* presented Merton with a dilemma. Lord Krishna with his devotee Arjuna stood on the battlefield of Kurukshetra. When Arjuna, reluctant to enter battle against his opponents—actually relatives and friends—wondered what his duty was, Lord Krishna counseled him to fight. This he ultimately did. Was the ultimate message of the *Gita* therefore the rightness of warfare? The great Indian nationalist Bal Gangadhar Tilak (1856–1920) thought so. Mahatma Gandhi denied it, preferring to interpret the *Gita* as a spiritual battle going on within each individual. Merton, consonant with his pacificism, convinced himself that the *Gita* does not justify all wars, but only wars subject to all kinds of restrictions, similar to the severe limitations on warfare in the Christian Middle Ages and in ancient India.[49] This takes a romantic view of both medieval Christian and traditional Indian warfare. As a further justification he added that the *Gita* those fulfillment of one's duty as a warrior as an example of the 'most repellent duty'.[50]

Closer to the truth is Merton's interpretations of the *Gita* as a divine counsel to fight the battle of life unselfishly in accordance with the divine will. It constituted a fusion of worship, action and contemplation. His ideal was the yogi who steadfastly concentrated on God and thereby rose above all sense perceptions. This is evident from the only lengthy quotation he made from the *Gita,* in the poetic rendering of Sir Edwin Arnold:

> Steadfast a lamp burns sheltered from the wind;
> Such is the likeness of the Yogi's mind
> Shut from the sense-storms and burning bright to Heaven.
> When mind broods placid, soothed with holy wont;
> When self contemplates self, and in itself
> Hath comfort; when it knows the nameless joy
> Beyond all scope of sense, revealed to soul—
> Only to soul! and, knowing, wavers not,
> True to the farther Truth; when, holding this,
> It deems no other treasure comparable,
> But harbored there, cannot be stirred or shook
> By any gravest grief, call that state 'peace',
> That happy severance Yoga; call that man
> The perfect Yogin![51]

For a lover of contemplation this passage must have held special fascination. But what attracted Merton above all to the *Gita* was its advocacy of the balanced life of contemplation and action represented by the karma yogi. Man ought to work selflessly, realizing that his work had cosmic significance. Was not, after all, God the supreme cosmic worker, artist and playwright, and the whole cosmic show a *lila* (play, sport)? The idea of the cosmic play, God's sport, appealed to him. Not by chance did he entitle the last chapter of *New Seeds of Contemplation* 'The General Dance': ' . . . the Lord plays and diverts himself in the garden of his creation . . . . '[52] Such a realization enables man to work joyously, free from anxiety, a willing instrument of the Lord. The ultimate goal is to attain total detachment from the results of action and even from the blissful effect of contemplation. Merton contrasted the attitude of the accomplished yogi with that of the average American who hectically works for material rewards.[53] That only few Hindus lived up to the lofty teachings of the *Gita* was, as we might expect, ignored by Merton.

In view of Merton's obvious admiration for some aspects of Hinduism, the question arises of how he viewed Christian missionary activity among Hindus. In a talk to the monks at Gethsemani in 1963 he expressed a point of view common among Christian missionaries since the end of the nineteenth century.[54] Hinduism ought to be respected but should be viewed as a stepping stone towards Christianity. Hindu teachings would find their fulfillment in Christianity. He took issue with the Hindu claim that their religion embraced all other religions, terming such an attitude too eclectic, a veritable mishmash.[55] He was by no means satisfied with Hindu willingness to accomodate Christianity by placing statues of Christ in their temples side by side with Hindu deities.

While Merton emphasized that 'we got the true faith,'[56] he condemned the efforts of missionaries to impose western civilization upon India, and he severely criticized Christian missionaries for their indiscriminate hostility to traditional Hindu customs. Just as he admired Jesuit attempts in China to accommodate the Chinese cultural heritage, he sympathized with the approach of the Italian Jesuit Father Robert de Nobili (1577–1656), who, believing strongly in the indianization of Christianity, had assumed the dress and life style of a *Brahman* (priest) and subsequently of a homeless wanderer. He had become a vegetarian, acquired a knowledge of Tamil, Telegu, and Sanskrit, and has been responsible for building churches according to Indian architectural design. As a result of various adaptations, Nobili had been eminently successful in converting Hindus to Christianity. Yet, in 1704, his reforms were disavowed by Rome and as a consequence proselytizing restricted.[57] Throughout his writings dealing with missionary efforts Merton continued to express his disapproval of attempts to identify Christianity with western civilization, above all, with a modern materialism he felt totally contrary to the teachings of Christ.[58]

Merton's most definitive statement on the relationship of Hinduism to Christianity can be found in a letter of 22 June 1965, to Philip L. Griggs (later known as Swami Yogeshananda). Griggs had accused Merton of being unfair towards Hinduism in his preface to Dom Denys Rutledge's book *In Search of a Yogi.*[59] Griggs had interpreted some of the statements of the preface to imply a spiritual inferiority of Hindu sages to Christian saints. Merton retorted:

> The Church's teachings on nearness to God is that he [who] loves God better, knows Him better, and is more perfectly obedient to His will, is closer to Him than others who may love, know, and obey Him less well. Since it is perfectly obvious that a Sadhu might well know God better and love Him better than a lukewarm Christian, I see no problem whatsoever about declaring that such a one is closer to Him and is even, by that fact, closer to Christ.[60]

Recognizing that comparing a sadhu to a lukewarm Christian would not convince Griggs that he was unbiased, Merton added that: 'Catholics believe that the Church does possess a clearer and more perfect esoteric doctrine and sacramental system which "objectively" ought to be more secure and reliable as a means for men to come to God and to save their souls'.[61] At the same time he admitted that God was not bound by any theological framework. The Holy Spirit could manifest himself more strongly in a Hindu monk than in himself (Merton). At the same time he affirmed a basic difference between Hinduism and Christianity: while Hindus believe man to be essentially, ontologically divine, the Catholic church taught that man was divine by grace alone and consequently can regard Hindu holy men as divine 'by adoption' only.[62]

That Merton, at least during the last phase of his life, had no enthusiasm for missionary enterprise among Hindus and instead preferred a dialogue on equal terms between Hindus and Christians can be deduced from his discussion of Gandhi's relationship to Christianity. Mahatma Gandhi he regarded as the ideal example of a non-Christian practising Christian principles. Already in 1931, when Merton was still a schoolboy at Oakham, Gandhi's visit to London to participate in the second Roundtable conference had been the occasion of Merton's defending Gandhi's views against a football captain who had shown no understanding for the 'naked fakir'. The young Merton sympathized with Gandhi's non-violent fight for India's independence. As his own spiritual development unfolded he saw in Gandhi a fellow traveller on the path towards a more spiritual and more harmonious world civilization based on a balance between action and contemplation and combining the best of East and West. Comparing his spiritual evolution with that of the Mahatma he was struck by the mysterious workings of East–West cultural inter-action and concluded: 'One of the most basic truths of our time is the mysterious fact that the full spiritual identity not only of cultures but of individual persons remains a secret gift that is in the possession of another. We do not find ourselves until, in meeting this other, we receive from him the gift, in part at least, to know ourselves.'[63] Merton had found his way to Christ through a Hindu monk; Gandhi had rediscovered his own tradition through exposure to Westerners and through study of the Gospel. In finding their tradition they had treasured in it all the values that were universal. Gandhi had accepted all 'in Christian thought that seemed relevant to him as a Hindu'.[64] Similarly, Merton throughout his eastern quest had accepted those values of the East that were relevant to him as a Catholic. In each case exposure to another culture had resulted in a deepening understanding of his own tradition, accompanied by an increased respect for the other's culture.

Merton was in wholehearted agreement with Gandhi's expectation that 'When the practice of *ahimsa* (non-violence) becomes universal, God will reign on earth as he does in heaven'.[65] While Merton's consistent opposition to violence was based on Christ's teachings, it undoubtedly drew inspiration from the Mahatma's living example. Merton was convinced that the Hindu Gandhi practised the teachings of the Gospel better than most 'genuine' Christians. In particular, he applauded the fact that Gandhi considered politics and social issues part of religion, avoiding the ominous separation of the sacred from the profane which typifies the modern West.[66] Most fervently Merton welcomed Gandhi's unequivocal condemnation of modern industrial society. He reserved special praise for Gandhi's struggle for the untouchables, for it was politically 'unwise', and carried out in a spirit of self-sacrifice. Using fasting and self-purification as a part of *satyagraha* (soul force), and aiming at overcoming the enemy through love appealed to Merton, who could not help but see a parallel between the plight of the untouchables in India and the situation of the Blacks in the United States. In both cases he thought it imperative to reconcile oppressor and oppressed. Gandhi and Merton knew full well that the oppressed were in no way more noble than the oppressors. Both were ego-enslaved and sinful. In this connection Merton refered to St Thomas Aquinas and Erasmus, who also stressed that evil could be overcome by self-suffering, by showing love rather than hatred for those who perpetrate evil.[67] Bringing about harmony in human relations required a radical change of heart, a *metanoia.* Gandhi's insistence on a conversion was also suggested to Merton by Coomaraswamy's article 'On Being in One's Right Mind', in which Coomaraswamy emphasized the need to repent, to become a new man by solving one's inner conflict between the divine and human element in our nature. Only then does a transformation ensue: 'from human thinking to divine understanding'.[68] And only then can one act on the basis of an inner freedom, in agreement with one's true self. Since the majority of Indians never fulfilled Gandhi's prerequisite for *satyagraha,* it had not been possible to test the principle under the best possible conditions. Clearly Gandhi and Merton were utopians who frequently ignored the human condition. In no other way can one explain their inability to acknowledge the failure of *satyagraha* which is shown by the massive outbreaks of violence between Hindus and Moslems that accompanied India's independence–1947.

Sympathy with Gandhi's aversion to modern western civilization led Merton to take sides with Gandhi against India's poet laureate, Rabindranath Tagore, on the question of boycotting western institutions, especially western schools. In contrast to the Mahatma, Tagore, although opposed to western exploitation of India, favored modern western civilization, believing it could be harmonized with India's traditional values. Merton, on the other hand, disapproved of modern western civilization and its vehicle, modern western education, which he termed 'barbaric'.[69]

Merton fittingly concluded his introduction to his book *Gandhi and Non-Violence* by drawing attention to Gandhi's message of love and truth and its accordance with Pope John XXIII's encyclical *Pacem in Terris,* which 'has the breadth and depth, the universality and tolerance of Gandhi's own peace-minded outlook'.[67]

The last phase of Merton's encounter with Hinduism took place on Indian soil, and is depicted, although sketchily, in *The Asian Journal.* Although, as we will point out, Tibetan Yoga and Buddhism in general received Merton's primary attention during his two months' stay in Asia, he did expand his understanding of Hinduism while there. Finally he was able to observe it at first hand rather than through books. It is highly significant that during his visit to Calcutta he tried to locate the ashram of Dr Brahmachari who had returned to India many years earlier.[71] Merton had hoped, although in vain, for one more meeting with the first Hindu who had influenced his life thirty years earlier. It would have been a memorable meeting. Alas Dr Brahmachari was living in East Pakistan by then and was not able to visit India because of strained relations between India and Pakistan.[72] Merton participated in Bengal's greatest religious festival, the worship of the goddess Kali, the fierce and bloody consort of Shiva, the destroyer. On that occasion clay images of Kali, black skinned, her tongue hanging out, wearing a garland of skulls around her neck, are displayed all over Calcutta. What must have gone on inside Merton's mind as he contemplated her dread appearance? In Madras he was taken to Kapaleeswara, the temple of Shiva: 'Extraordinary life and seeming confusion of the temple, full of people milling around barefoot (I too) in the sand, children playing and yelling, dozens of shrines with different devotions going on . . . . '[73] Was this the Hinduism of the Upanishads and of the *Bhagavad Gita*? While in Madras Merton had an opportunity to meet Dr V. Raghavan, Professor-emeritus of Sanskrit at the University of Madras. Their conversation turned to Indian aesthetics and its relation to religious experience. Merton alluded to William Blake whose views on art and religion were so strikingly similar to those of the Hindus. The knowledge which he had gained while writing his M.A. dissertation thirty years earlier proved unexpectedly useful!

While experiencing cultic Hinduism 'in the flesh', Merton spent considerable time on his Asian journey reading further works on various aspects of Hinduism. He relished the poetry in praise of Krishna and his beloved consort Radha. In that poetry a different aspect of Lord Krishna is depicted from that presented in the *Bhagavad Gita.* Here the erotic exploits of the charming cowherd-god Krishna, especially his love for the beautiful milkmaid Radha, are commemorated. The erotic encounters of Krishna with Radha and other milkmaids are usually interpreted as symbolic of the love of God for the human soul, reminding one of similar interpretations of the Song of Songs by St. Bernard of Clairvaux and St. John of the Cross.[74] Merton also studied the life of the eleventh century qualified monist, Ramanuja. In contrast to the absolute monist, Sankara, Ramanuja rejected the view that God and man were one. Instead he maintained that even though man could attain a close union with God, the two remained distinct; God as well as man retained his individuality.[75] Although Merton knew that the bridal mysticism of Krishna and Radha and the qualified monism of Ramanuja were closer than absolute monism to Christian tradition—since they made allowance for a personal deity—he seems to have been more attracted to Sankara (ninth century A.D.). *The Asian Journal* contains copious quotations from Sankara's *Crest-Jewel of Discrimination.* Was it Sankara's bold denial of ultimate reality to the phenomenal universe that intrigued Merton? The passages he quoted focus on the need to rise above name, race, and form by means of discrimination and to realize that One only exists, and 'That is Thou'.

Merton appears to have had further plans for the exploration of the many faces of Hinduism. Ed Rice, his friend from student days at Columbia, relates that Merton's future reading assignment was tantric Hinduism, an unorthodox form of rituals including magic and sexual symbolism.[76] Since *The Asian Journal* is replete with references to tantric Buddhism, it is very likely that Merton also wanted to familiarize himself with the tantric dimension of Hinduism. Obviously his eagerness to investigate mankind's religious heritage had remained undiminished, even increased!

Chapter IV
MERTON IN ASIA

WITH ALL HIS LOVE for Asia it is understandable that Merton was eager to see the continent of his dreams and hopes, to come face to face with that Asia which he had so far known through books and a few contacts with Asians living in the West. The first concrete possibility to visit Asia presented itself in 1964, when Heinrich Dumoulin invited him to spend some time in a Trappist monastery, Our Lady of the Lighthouse, in Hokkaido. Dumoulin was in agreement with Merton that Christian monasticism could benefit from such an exposure to Far Eastern spirituality. He also believed that Merton's literary activity would benefit from an Asian encounter. Above all it would 'give him a competence of judgement that cannot be acquired by reading books'.[1] Dumoulin mentioned that some Catholic theologians were not sufficiently respected in the East precisely because they lacked a personal experience of Asian religious life. Merton naturally was only too willing to go to Hokkaido. He was especially intent upon participating in Zen retreats with a view to deepening his contemplative experience and at the same time engage in a dialogue with Zen monks.[2] Although Dumoulin, at Merton's suggestion, wrote to Dom Ignace Gillet, Abbot General of the Trappist Order in Rome, to intervene in behalf of Merton, his request was rejected by his abbot, Dom James Fox, who considered it 'completely inconsistent with the contemplative life'.[3] Merton was profoundly disappointed:

> . . . . I naturally renounce it willingly and without after thought. What troubles me more is the failure of understanding, which I seem to be able less easily to forget . . . I must admit it is rather wounding to be told that such a project is 'not from God.' Paradoxically, it is Zen itself which gives the most practical perspective by which to see and to accept all this![4]

Four years later Merton's yearning to visit Asia finally found fulfillment. A Benedictine group had made arrangements for convening a conference of Asian monastic leaders in the middle of December 1968, in Bangkok, Thailand. Dom Jean Leclercq, a Benedictine scholar with whom Merton has corresponded for several years, saw to it that Merton was scheduled to give one of the major addresses at the conference. Abbot James Fox had just resigned. The new abbot, Flavian Burns, not only gave him permission to go to Bangkok but also allowed him to visit various Christian and Buddhist monasteries in different Asian countries and to participate in a Spiritual Summit Conference organized by the Temple of Understanding in Calcutta.[5]

'The length of my stay in Asia is indeterminate,'[6] Merton advised his friends in a circular letter written in September 1968. He was 'absolutely bouncing with expectation',[7] for he would 'drink from the ancient source of monastic vision'.[8] Although he reemphasized, as befits a true mystic, 'Our real journey is interior',[9] he had a craving to get away from Gethsemani. As much as he channeled his need for new experiences through a wide variety of readings and occasional visits to Louisville, he was thirsting for an outward journey after more than twenty-six years of monastic life, and 'after years of waiting and wondering and fooling around'.[10] Facetiously he had written a year earlier to his friend Seymour Freedgood: 'Not that I want to spend the rest of my life in a kindergarten . . . . '[11] He personally did not see any conflict between an inner and an outer pilgrimage. One cannot disagree with his argument that the inner and outer pilgrimage can reinforce each other. And yet one cannot avoid the conclusion that something within Merton had remained unfulfilled. Twenty-six years of monasticism had not given him the 'peace that passeth understanding'. A certain sense of incompleteness is indicated by his observation: 'May I not come back without having settled the great affair. And found also the great compassion, *mahakaruna.*'[12] Since Merton used *mahakaruna* to signify Absolute Reality, 'oneness with the other in Christ',[13] one may interpret this to signify that he hoped to find in Asia a more intimate union with God than he had experienced so far. This would agree with Abbot Flavian Burns' observation that Merton 'undertook this trip to Asia in the same quest for God. His letters to me from there were buoyant with hope for further progress in his quest.'[14]

'Buoyant with hope' is an appropriate description of Merton's mood upon departure from San Francisco eastward, bound on 15 October 1968: 'The moment of take-off was ecstatic . . . . I am going home, to the home I have never been in this body . . . . '[15] And on the flight from Honolulu to Bangkok via Tokyo and Hong Kong he declared exuberantly: 'When the stewardess began the routine announcement in Chinese I thought I was hearing the language of Heaven'.[16]

In general he greatly enjoyed his pilgrimage. He relished arrack, whiskey and bloody marys and delighted in observing and talking to beautiful women. Embassy parties thrilled him.[17] What welcome relief after over twenty-six years of monastic austerity! The 'fun' aspect of his journey must not be overly stressed, though. The author-monk continued his rigorous intellectual pursuits during the whole trip. While travelling and intensely absorbing new sites and engaging in religious dialogue with natives of India, Sri Lanka, Tibet, and Thailand, Merton was engaged in a heavy reading schedule. In fact, he had to pay for excess baggage for carrying with him an extra load of books.[18] The bulk of his reading centered on Asian religions and culture including Marco Pallis, *Peaks and Lamas; The Way and the Mountain,* W.Y. Evans-Wentz, *The Tibetan Book of the Dead; Tibetan Yoga and Secret Doctrine, Tibet's Great Yogi Milarepa.* But it is indicative of the continued catholicity of his interests that he found time during his Asian journey to read, among others, D.H. Lawrence, *Twilight in Italy,* Hermann Hesse's *Steppenwolf* and Anias Nin, *A Glass Bell.*[19]

In a summary of Merton's Asian experiences it is essential to discuss his encounters with Asian people, the exponents of eastern wisdom with whom he had longed to dialogue. The first significant Asian he met upon arrival in Thailand was the monk Phra Kantipalo who spoke to him about 'mindfulness'. Kantipalo explained that, 'Mindfulness is the awareness of what one is doing while one is doing it and of nothing else'.[20] In other words, mindfulness signifies total concentration on one's present task without yielding to mental distractions. Without mindfulness meditation becomes a farce. To illustrate how mindfulness differs from ordinary consciousness, Kantipalo cited a Zen story, for Zen Buddhism like Theravada Buddhism in Thailand was based on mindfulness. When a Zen master was asked to explain the essence of his teaching, he replied: 'When hungry I eat; when tired I sleep.' This reply did not satisfy the questioner. So the

Zen Master added: 'When they eat most people think a thousand thoughts; when they sleep they dream a thousand dreams.'[21]

More important than his contacts with Thai monastics were Merton's encounters with Tibetans whom he met in northern India. They 'have a peculiar intentness, energy, silence, and also humor'.[22] At Ghoom near Darjeeling he dialogued with Chatral Rimpoche, a renowned monk who had meditated in seclusion for thirty years. They discussed the goals of Christianity and Buddhism, which led them to focus on *dzogchen*: ultimate perfect emptiness 'beyond God'.[23] Merton felt in total harmony with Chatral: 'We were somehow *on the edge* of great realization and knew it . . . . '[24] Was Merton flattered when Chatral called him a 'natural Buddha' or assured him that both he and Chatral were on the verge of Buddhahood, nay, might attain Buddhahood in this life?[25] Were they bantering each other? Although Merton gave the impression that he considered himself Chatral's spiritual equal, he also indicated that he might choose Chatral as his guru. But he added that he was not so sure that he was in need of a guru.[26]

It seems that Merton seriously considered studying Tibetan meditation techniques under a guru, for he brought this subject up repeatedly. While discussing Tantrism with a Nyingmapa (Order of the Ancients) monk, Merton expressed interest in Tantric initiation.[27] During Merton's interviews with the Dalai Lama he was told about a Tibetan teacher at the University of Wisconsin and one in Rikon, Switzerland—both highly recommended to him by the Dalai Lama. The Dalai Lama also urged him to study *Madhyamika,* a Buddhist school of philosophy founded by Nagarjuna (ca. 150–250 A.D.). Merton and the Dalai Lama exchanged views on the nature of ultimate truth. In the end the Dalai Lama is said to have called Merton a 'Catholic *geshe*' (a learned Lama, equivalent to a Doctor of Divinity).[28] One further interview was of significance: the meeting with the Dalai Lama's private chaplain, the Khempo of Namgyal Tra-Tsang. The Khempo evaded the discussion of metaphysics. Instead he emphasized the prior need to experience absolute compassion for all beings before seeking enlightenment. He too urged Merton to seek a guru.[29] Summing up his impression of Tibetan spiritual masters Merton concluded that some of them had attained greater heights in meditation than Catholic contemplatives. Yet from a practical point of view he did not consider Tibetan meditation methods suitable for the West. They appeared to him too complicated. He continued to favor Zen as a more appropriate way for Christian contemplatives and voiced his hope that he would encounter Zen monks later during his Asian journey.[30]

As a direct outcome of Merton's meetings with Tibetans he further immersed himself in Buddhist studies. He spent some time examining Guiseppe Tucci's *The Theory and Practice of the Mandala* (London, 1961). He eventually gained an intellectual understanding of the mandala, a symbolic representation of the cosmos upon which one is to focus to attain an integrated view of the outer and inner universe. But centering upon a mandala did not suit him: ' . . . I have a sense that all this mandala business is, for me, at least, useless.'[31] In contrast he was enthusiastic about *Madhyamika* philosophy, which he investigated on the basis of T. R. Murti's *The Central Philosophy of Buddhism: A study of the Madhyamika System* (London, 1955). Copious quotes from Murti's work throughout *The Asian Journal* clearly reveal why *Madhyamika* appealed to Merton. The *Madhyamika* philosophy, similar to Zen, aimed at liberating man from conventional ways of thinking. It emphasized the need to transcend all points of view, all concepts. There was an underlying assumption that reality would surface once all points of view had been rejected. That highest reality, referred to as *Prajnaparamita* (Wisdom gone beyond) was 'transcendent to thought, non-relative, non-determinate, non-discursive, non-dual'.[32] Writing to his abbot on 9 November 1968, Merton reported about his study of *Madhyamika* 'which is not speculative and abstract but very concrete and fits in with the kind of sweeping purification from conceptual thought which is essential'.[33]

Apart from dialoguing with individual Asians Merton made an attempt to reach some overall conclusions about Asia. In less than two months in Asia, he did not have sufficient time to arrive at definitive judgements about the vast and infinitely varied continent. Even during his short stay he felt it necessary to revise some of his views. This applied especially to Calcutta, Delhi, and Bangkok which he visited twice:

> It is good to have a second time around with these cities. Calcutta, Delhi, and now Bangkok. It now seems a quite different city. I did not recognize the road in front of the airport, and the city which had seemed, before, somewhat squalid, now appears to be, as it is, in many ways affluent and splendid. What has happened, of course, is that the experience of places like Calcutta and Pathankot has changed everything and given a better perspective to view Bangkok . . .[34]

The overriding question was, what was the real Asia, the Asia uncorrupted by the West, the Asia of his dreams? By searching for the 'real Asia' Merton again reveals the dilemma of the utopian who defines reality according to his ideals. Moreover, thinking in terms of one Asia is futile in view of the marked differences between North Asian, Central Asian, South Asian, Southeast Asian, and East Asian civilizations. While visiting the big cities Merton was made painfully aware of western influences. The passport officers in Bangkok disgusted him: 'Fascist faces . . . . Tired, crafty, venal faces, without compassion . . . . Men worn out by a dirty system.'[35] In Calcutta he was repelled by movie posters which exhibited the worst features of East and West, the perversion of eastern culture by American advertising techniques.[36] It was, however, not America but Great Britain, the standard bearer of western civilization in India for more than two hundred years, which he felt had caused the major harm to Indian culture. Merton was clearly irritated by the continued English presence in India. While staying at the Mim Tea Estate above Darjeeling, probably because he was suffering at that time from a bad cold, he questioned whether it had been really worthwhile travelling to Asia:

> English hats, tweeds, walking sticks, old school ties (St. Joseph's) . . . for the rich ones at least. Shivering in the Windamere [Hotel] over Madhyamika dialectic . . . is that the "real Asia"? I have a definite feeling it is a waste of time—something I didn't need to do. However, if I have discovered I didn't need to do it, it has not been a waste of time.[37]

Of all the cities of India, Calcutta impressed and baffled him most, for in Calcutta, capital of British India until 1911, the strange incongruity of the mingling of East and West was acutely conspicuous. Poverty, filth, political extremism, the highest cultural sophistication—that is Calcutta. Comparing the madness of Calcutta with that of America he concluded:

> Calcutta is shocking because it is all of a sudden a totally different kind of madness, the reverse of that other madness, the mad rationality of affluence and over-population. America seems to make sense, and is hung up on its madness, now really exploding. Calcutta has the lucidity of despair, of absolute confusion, of vitality helpless to cope with itself.[38]

On his second visit to Calcutta he learned to love the city: ' . . . the subtle beauty of all the suburban ponds and groves, with men solemnly bathing in the early morning and white cranes standing lovely and still amid the lotuses . . . . '[39]

Merton was entranced by the mighty beauty of the Himalayas and charmed by the loveliness of Madras and of South India in general.[40] But of all the sights in India, Mahabalipuram, an abandoned temple city south of Madras, impressed him most favorably. Here at last was the Asia of his dreams. It is significant that he found his dream embodied in an abandoned city, belonging to the past, a past which he had not experienced. That the Pallava dynasty responsible for the magnificent architecture of Mahabalipuram had been chronically engaged in violent wars was probably not known to Merton.[41] Or was it conveniently ignored?

It was another ruined temple city, Polonnaruwa, in Sri Lanka, which produced the peak experience of Merton's Asian journey. Contemplating the mighty Buddha statues apparently brought him to the goal of his pilgrimage, of his quest for meaningfulness:

> . . . Then the silence of the extraordinary faces. The great smiles. Huge and yet subtle. Filled with every possibility, questioning nothing, knowing everything, rejecting nothing, the peace not of emotional resignation but of Madhyamika, of sunyata, that has seen through every question without trying to discredit anyone or anything—without refutation—without establishing some other argument . . . . The thing about all this is that there is no puzzle, no problem, and really no 'mystery'. All problems are resolved and everything is clear, simply, because what matters is clear. The rock, all matter, all life, is charged with dharmakaya [the ultimate body of the Buddha] . . . everything is emptiness and everything is compassion. I don't know when in my life I have ever had such a sense of beauty and spiritual validity running together in one aesthetic illumination. Surely, with Mahabalipuran and Polonnaruwa my Asian pilgrimage has come clear and purified itself. I mean, I don't know what else remains but I have now seen and have pierced through the surface . . . . This is Asia in its purity, not covered over with garbage, Asian or European or American, and it is clear, pure and complete . . . .[42]

The date of this mystic experience was December 3, one week before Merton's death; his statement: 'I don't know what else remains' assumes special significance. He had found the great compassion, *mahakaruna,* which prompted him to proclaim the interdependence of all beings during his last talk in Bangkok, two hours before his earthly exit. For a Christian monk to couch his spiritual attainment in Buddhist terminology might indicate that he had concluded there was a point of convergence of Christian and Buddhist experience. At the same time he was sufficiently cautious to focus on the natural and not the supernatural significance of Buddhism and Christianity.

Since Merton considered Polonnaruwa representative of Asia in its purity, it is only fair to ask whether he would have retained this view had he known that at the time when the Buddha statues had been constructed, Sri Lanka had in no way been exempt from internal and external strife. Had Merton lived in ancient or medieval Sri Lanka he might have found fault with its civilization, too. What Merton never could bring himself to admit was that an individual awareness of *mahakaruna* did not guarantee socio-economic and political harmony; that there was no golden age in the ancient world of Asia or of Europe.

What general insights had Merton attained during his Asian journey? This is a legitimate question, for Merton had not only wanted to benefit personally from exposure to Asia but had also hoped to contribute to the renewal of Christian monasticism and Western society. He became convinced that compassion had to be combined with detachment and wisdom. Only through wisdom could selfishness be destroyed and only thereafter was unconditional, selfless love feasible.[43] Likewise, wisdom could only be attained following the practice of contemplation in solitude. His Asian experience reinforced his belief that a monastic dialogue was imperative. He conceded that there were marked differences between the great religions on the doctrinal level, differences which could not be erased, yet he reiterated his conviction that there were great similarities in the area of religious experience.[44] A dialogue need not threaten each participant, because a solidly grounded Christian would not be tempted to abandon his faith while profiting from Buddhist or Hindu techniques. He argued that an interreligious vocabulary was developing. But that was an optimistic estimate. William Johnston's view that such an interreligious vocabulary will be worked out in the next century[45] appears to me more realistic. But more important than any verbal dialogue was communication on the 'preverbal' and 'postverbal' level, Merton averred. After all, spoken words had intrinsic limitations. By 'preverbal' level he meant 'the "predisposition" of the mind and heart necessary for all "monastic" experience whatever'.[46] The 'preverbal' level presupposes spontaneity, freedom from conventional thinking. It enables one to communicate with someone representing an alien tradition and discover 'a common ground of verbal understanding with him'.[47] From this point he hoped the two communicants would be able to interact with one another in silence on an experiential level, namely the 'postverbal' level.[48] To avoid theological wrangling, the most fruitful 'dialogue' would have to be conducted on both sides by monastics with many years experience in silent contemplation. At the same time Merton warned against the dangers of syncretism.[49] A genuine dialogue would lead to 'the growth of a truly universal consciousness . . . of transcendent freedom and vision . . . '[50] and ultimately he expected: ' . . . what one might call typically "Asian" conditions of nonhurrying and patient waiting must prevail over the Western passion for immediate visible results.'[51]

Through his study of *Madhyamika* as well as through his prior exposure to Zen he had become accustomed to seeking to transcend structures, both governmental (secular) and monastic, in order to attain 'perfect freedom'. His need to go beyond structure was strengthened by listening to the experiences of Tibetan monastic refugees. One monk told him that when they had to leave their monastery and their homeland in the wake of the Chinese Communist invasion, they were left without the accustomed structure on which they had based their lives and some even their faith. Before their departure, the monk had asked his abbot for guidance, only to be told that he was on his own.[52] In the end that is what all of us must be ready to do, Merton realized. One could never know when the institutional props might be removed, when we might have to fend for ourselves, guided by an inner awareness which results from silent contemplation. He concluded that the monastic structure, as any other structure, served only one purpose: enabling the individual to find that inner center, that true self which alone guarantee true freedom.

In his last Bangkok address, 'Marxism and Monastic Perspectives,' on 10 December 1968, Merton clearly stated that he had transcended all structures, all division, whether national, racial, cultural, or even religious. Paraphrasing St Paul, he declared: 'There is no longer Asian or European for the Christian . . . . So you respect the plurality of these things, but you do not make them ends in themselves . . . we accept the division, we work with the division and we go beyond the division.'[53] One attains the freedom to go beyond all divisions among human beings, all living beings, through *mahakaruna,* the great compassion. And monasticism must consider it its sacred task to enable monks to experience *mahakaruna.* Only then has a monastic renewal been successfully effected.

In his enthusiastic effort to encompass everything, Merton also attempted to make room for Marxism. To explain why he continued to be attracted to Marxism, one has to consider that throughout his life he had vacillated between solving the world's problems through individual conversion—the spiritual approach—and changing the socio-economic organization of society—the Marxian approach. Ideally he would have wanted to combine the two approaches and therefore he eagerly favored a Catholic-Marxian dialogue. In support of his point of view he cited the French Marxist Roger Garaudy, who had become interested in St Teresa of Avila and in Christian monasticism in general.

Meeting with Tibetan refugees ought to have dampened his enthusiasm for Marxism, for the Tibetans gave living testimony to the Chinese communists' ruthless persecution of Tibetan monastics; their efforts to wipe out all traces of religion in Tibet constitute a dark page in the history of humanity. In preparation for his talk on the relationship of Marxism to monasticism on December 10, Merton had discussed the subject with several Tibetans. One Tibetan monk, Sonam Kazi, who had served as an official interpreter for the Dalai Lama in negotiations with the Chinese, informed Merton that unless one had wealth, one need not fear the Communists. And he indicated that he believed that many Tibetan monks had been too affluent.[54] On the other hand, the Dalai Lama emphasized that Communists and monks could not cooperate as long as the Communists made totalitarian claims and did not confine their actions to the socio-economic realm.[55] By the time Merton delivered his talk he had received sufficient proof of Communist atrocities in Tibet that he found it necessary to state:

> This is *not* a talk addressed to the needs of those brethren here present who have been in totally different and much more existential contact with Marxism, namely, those who have had to flee for their lives from Communist countries. This is not the problem I am talking about, the problem of life and death, when for bare survival one simply has to get away from an enemy who seeks to destroy or completely convert one. I don't see that there is much that can be said about this. . . . I am thinking much more in terms of the kind of Marxist thought that influences the youth of the West and will possibly influence some of the youth of Asia —and that will, I think, be influential with the kind of people we could really be in vital contact with in the East, that is to say, intellectuals.[56]

Here obviously Merton identified with intellectuals who prefer to look at Marxist theory and to ignore the reality of applied Marxism as practised by communist regimes both in

Europe and in Asia. Merton failed to see the incongruity of his approach, which ran counter to his usual insistence on coming to grips with reality, his existentialist way of looking at religions. The existential contact with Tibetan refugees he conveniently set aside to make room for a strictly theoretical discussion.

In his lecture on 'Marxism and Monastic Perspectives' Merton acknowledged his indebtedness to the Neo-Marxist Herbert Marcuse, who incidentally never visited the U.S.S.R. or any other communist country. Merton categorized Marcuse as a monastic thinker.[57] He defended this amazing categorization: 'The monk is essentially someone who takes up a critical attitude toward the world and its structure ....'[58] And he added that Marxists and Christians shared the desire to move from '*cupiditas* to *caritas*',[59] and both had in common the aim: 'from each according to his ability, to each according to his need'.[60] Referring to Marcuse's book *One Dimensional Man,* Merton declared his agreement with Marcuse's thesis that highly technologically organized societies, as the United States and the Soviet Union, tend to become totalitarian. That agrarian China, especially during the Ch'in dynasty (221–206 B.C.) had experienced totalitarian rule he conveniently overlooked. Merton also shared Marcuse's concern about a potential atomic holocaust. But Merton also ignored the fact that Marcuse, in agreement with his guru Marx, totally disregarded man's spiritual dimension—an indication of the incompatibility of Marxism and Christianity. Nor did he mention that the United States provided for spiritual freedom, including the right to embrace or not embrace monasticism, in contrast to the anti-monastic regulations in the Soviet Union. Most importantly, Merton overlooked that Marcuse's solution for the problems of 'One Dimensional Man' called for a revolution of 'outcastes and outsiders' against the existing structure. Marcuse did not come to grips with man's inherent imperfection, his sinfulness, an article of faith in Christianity.

Towards the end of his lecture, however, Merton did re-emphasize the spiritual solution for mankind's ills, the solution that had led him to Gethsemani and had made it possible for him to experience the great compassion seven days before his talk. Through his experience of the great compassion he had attained a perspective where divisions no longer mattered. With that vision in mind he could express his faith in a state of liberty 'that nobody can touch, that nobody can affect, that no political change of circumstance can do anything to'.[61] He admitted that that sounded 'a bit idealistic. I have not attempted to see how this works in a concentration camp, and I hope I will never have the opportunity . . . . '[62] Merton might have cited the example of St Maximilian Kolbe, the Polish priest who gave an inspiring testimony to the triumph of spirit over brute matter, by dying in a concentration camp, singing to the end to the glory of Virgin Mary.[63]

Merton concluded his talk by paying homage to eternal monasticism and by reaffirming his faith in a continuation of the dialogue between East and West: 'And I believe that by openness to Buddhism, to Hinduism, and to these great Asian traditions, we stand a wonderful chance of learning more about the potentiality of our own traditions . . . . '[64] A few hours later he was found dead, apparently electrocuted by a fan.

EPILOGUE

DURING HIS STAY in Dharamsala Merton recorded a dream in which he found himself back in Gethsemani in the habit of a Buddhist monk. Was he concerned about readjusting to Gethsemani after his immersion in Asian culture? Would he have wanted to live as a hermit somewhere else, where he might have had greater solitude? Did he really want uninterrupted solitude or did he not have a great need, at least intermittently, to enjoy human companionship? Even in Gethsemani, when after long years of pleading, he had finally, in 1965, been permitted to retire to a hermitage, Dom John Eudes Bambergerger, a fellow Trappist at Gethsemani, reported: 'In spite of his deep, unvarying, and intense attraction to solitude Fr Louis [Merton's monastic name] was one of the most sociable of men, who had an absolute need for human society . . . .'[1] In the *Asian Journal* Merton mentions that he still hoped to visit the Alps, Alaska, Scotland, Wales, and Switzerland.[2] Even after the great experience at Polonnaruwa when he declared that he did not know what else remained for him to do, he noted: 'But the journey is only begun. Some of the places I really wanted to see from the beginning have not yet been touched.'[3] There was something of a Faustian urge in Merton which was in conflict with his contemplative tendency. He had a 'western' craving for more and more experiences, for further inner and outer journeys. One wonders whether he had found that perfect balance between contemplation and activity which was his ideal. This in turn raises questions concerning the validity of his ideal. The ideal Asia was clearly not the contemporary Asia, just as the ideal Europe was in his estimation the Europe of the Middle Ages. It must have caused conflict within him that he did not see his ideal incarnated anywhere in his own age. While he had, at least momentarily, gained a vision of an ultimate harmony, he had obviously not found a way of applying a personal experience of harmony to society at large. His suggested fusion of ideal Marxism and monasticism constituted a very questionable solution.

Certain reservations concerning Merton's idealism notwithstanding, one cannot but be impressed by his steadfast search for God through contemplation and action inside and outside the monastery, within and beyond the western tradition. One also cannot help being deeply impressed by his genuine appreciation of Asian people and their religions. Christian dialogue with Asia will be continued and will even be intensified. Probably the most urgent and convincing message he bequeathed us is to be found in his last speech: 'There is no longer Asian or European for the Christian'. In an age of strife and the threat of nuclear annihilation we need to be reminded of our common humanity; to know that we are all God's children. It was for this reason that

Merton enthusiastically welcomed Vatican II's encouragement of a dialogue with non-Christians and its emphasis on 'what men have in common and what tends to promote fellowship among them'.[4] Over and over again Merton had emphasized the universal, those aspects that bind us together rather than those that separate us. It was for this reason that Merton concentrated on comparative spiritual theology rather than comparative dogmatic theology. Merton fervently urged us to listen to the voice of Asia: 'It is true that the visible Church alone has the official mission to sanctify and teach all nations, but no man knows that the stranger he meets . . . is not already an invisible member of Christ and perhaps one who has some providential or prophetic message to utter'.[5] While Merton was dedicated to dialoguing with non-christians, he retained his allegiance to the Catholic Church. It was in fact only because he was firmly grounded in his faith that he dared to explore other religious experiences which could only further expand his catholicity. Merton's life had been a testimony to the great truth which he had proclaimed: ' . . . the mysterious fact that the full spiritual identity not only of cultures but of individual persons remains a secret gift that is in the possession of another. We do not find ourselves until, in meeting the other, we receive from him the gift, in part at least, to know ourselves.'[6] It was Asia that bestowed, at least in part, the gift of self-understanding upon Thomas Merton. In return he presented the best of Asia to the West.

NOTES

Preface

1. Brother Patrick Hart, ed., *Thomas Merton, Monk: A Monastic Tribute* (Garden City: Doubleday Image Books, 1976), p. 220. New, enlarged edition (Kalamazoo: Cistercian Publications, 1983), p. 220.

Chapter One

1. Thomas Merton, *The Seven Storey Mountain* (New York: Harcourt, Brace 1948), p. 32.
2. *Ibid.,* p. 37.
3. *Ibid.,* p. 85.
4. Thomas Merton, *Emblems of a Season of Fury* (New York: New Directions, 1963), p. 73.
5. Thomas Merton, *Conjectures of a Guilty Bystander* (Garden City: Doubleday, 1966), p. 62; Thomas Merton, *Contemplation in a World of Action* (Garden City: Doubleday Image, 1973), p. 49.
6. *The Seven Storey Mountain,* p. 133.
7. Thomas Merton, *Seeds of Destruction* (New York: MacMillan, 1967), p. 84.
8. *Emblems of a Season of Fury,* pp. 77-78.
9. 'The Inner Experience: Notes on Contemplation,' (fourth draft), p. 29.
10. Thomas Merton to Archbishop Yu Pin, February 3, 1961; unpublished letter in the Thomas Merton Studies Center.
11. Thomas Merton to Marco Pallis, n.d.; unpublished letter in the Thomas Merton Studies Center.
12. Aldous Huxley, *Ends and Means: An Inquiry into the Nature of Ideals and into the Methods Employed for Their Realization* (New York: Harper, 1937), pp. 336, 381.
13. *Ibid.,* p. 381.
14. *The Seven Storey Mountain,* p. 187.
15. *Ibid.,* pp. 187-88.

16. *Ibid.*, p. 198.
17. Mohandas K. Gandhi, *An Autobiography: The Story of My Experiments with Truth* (Boston: Beacon, 1957), pp. 67, 90.
18. 'Author's Preface to the Japanese Edition,' in *Nanae Yo Yama (The Seven Story Mountain)*, trans. Tadishi Judo (Tokyo: Chuo Shupansha, 1966), pp. 11-12.
19. Thomas Merton, *The Waters of Siloe* (New York: Harcourt Brace, 1949), p. 3.
20. Thomas Merton, *The Sign of Jonas* (New York: Harcourt Brace, 1956), pp. 197-98.
21. *Ibid.*, p. 243.
22. Thomas Merton, *The Ascent to Truth* (New York: Harcourt Brace, 1951), p. 26.
23. Thomas Merton, trans., *The Wisdom of the Desert: Sayings of the Desert Fathers of the Fourth Century* (New York: New Directions, 1960), p. 9.
24. *Ibid.*, p. 5.
25. *Ibid.*, pp. 7-8.
26. *Conjectures of a Guilty Bystander*, p. 140.
27. *The Wisdom of the Desert*, p. 10.
28. *Ibid.*, p. 11.
29. Thomas Merton to Archbishop Yu Pin, February 3, 1961; unpublished letter in the Thomas Merton Studies Center.
30. Letter by Father Paul Chan, Secretary General Sino-American Amity, Inc., New York, N.Y., February 8, 1961 to Thomas Merton; unpublished letter in the Thomas Merton Studies Center.
31. Letter by John C. H. Wu to Thomas Merton, March 20, 1961; unpublished letter in the Thomas Merton Studies Center.
32. John C. H. Wu, *Chinese Humanism and Christian Spirituality* (Jamaica, New York: St. John's University, 1965), pp. 97-103.
33. Thomas Merton, 'A Christian Looks at Zen', introduction to John C. Wu, *The Golden Age of Zen* (Taipei, Taiwan: The National War College, 1967), p. 1.
34. Letter by John C. H. Wu to Thomas Merton, Good Friday, 1961: unpublished letter in the Thomas Merton Studies Center.
35. Letter by John C. H. Wu to Thomas Merton, April 1, 1961; unpublished letter in the Thomas Merton Studies Center.
36. Thomas Merton to John C. H. Wu, December 28, 1965; unpublished letter in the Thomas Merton Studies Center.
37. Thomas Merton, *New Seeds of Contemplation* (New York: New Directions, 1961), pp. 64-69.
38. A. L. Basham, *The Wonder that was India* (New York: Grove

Press, 1959), pp. 118-36.

39. Thomas Merton, *Mystics and Zen Masters* (1967; rpt. New York: Dell, 1978), p. 81.
40. *Ibid.,* p. 83.
41. *Ibid.,* p. 88.
41b. *Ibid.*
42. *Ibid.,* p. 46.
43. Thomas Merton to Archbishop Yu Pin, February 3, 1961: unpublished letter in the Thomas Merton Studies Center.
44. Thomas Merton, *Opening the Bible* (Collegeville, MN: Liturgical Press–London: G. Allen and Unwin, 1972), p. 52.
45. *Mystics and Zen Masters,* p. 60.
46. Julia Ching, *Confucianism and Christianity: A Comparative Study* (Tokyo: Kodansha International/Sophia University, 1977), p. 178.
47. *Ibid.* p. 216.
48. *Mystics and Zen Masters,* pp. 78-79.
49. *Ibid.,* p. 79.
50. *Ibid.,* pp. 53-56.
51. *Ibid.,* pp. 48-49, 75; Thomas Merton, *The Way of Chuang Tzu* (New York: New Directions, 1965), pp. 20-21.
52. *Mystics and Zen Masters,* p. 76.
53. *Ibid.,* p. 76.
54. *Ibid.,* p. 77.
55. *The Way of Chuang Tzu,* p. 10.
56. *Ibid.,* pp. 9-10.
57. *Ibid.,* p. 9.
58. *Ibid.,* pp. 9-10.
59. Hart, *Thomas Merton, Monk: A Monastic Tribute,* p. 51; *The Way of Chuang Tzu,* p. 9.
60. Hart, *Thomas Merton, Monk: A Monastic Tribute,* p. 51.
61. *The Way of Chuang Tzu,* p. 30.
62. *The Way of Chuang Tzu,* p. 30.
63. Thomas P. McDonnell, ed., *A Thomas Merton Reader,* rev. ed. (Garden City: Doubleday Image, 1974), p. 16.
64. *The Way of Chuang Tzu,* p. 88.
65. *Ibid.,* p. 10.
66. *Ibid.,* p. 99.
67. *Ibid.,* p. 69.
68. *Ibid.,* p. 153.
69. *Ibid.,* pp. 52-43.
70. *Ibid.,* pp. 11, 15; see also John Blofeld, trans., *The Zen Teaching of Huang Po on the Translation of Mind* (London: Rider, 1958)

and Charles Luk, ed. and trans., *The Transmission of the Mind Outside the Teaching* (New York: Grove Press, 1974).

Chapter Two

1. John Moffit, *Journey to Gorakhpur* (New York: Holt, Rinehart and Winston, 1972), p. 275.
2. Thomas Merton, *Zen and the Birds of Appetite* (New York: New Direction, 1968), p. 32.
3. *Ibid.*, pp. 30-31.
4. *Ibid.*, pp. 3-4.
5. *Ibid.*, p. 4.
6. Wu, *The Golden Age of Zen*, p. 2.
7. Letter by Father Heinrich Dumoulin to Thomas Merton, September 19, 1964; unpublished letter in the Thomas Merton Studies Center.
8. Heinrich Dumoulin, SJ, *A History of Zen Buddhism* (Boston: Beacon, 1963), p. 33.
9. *Ibid.*, pp. 61-63.
10. *Zen and the Birds of Appetite*, pp. 62, 78, 138.
11. Letter by Father Heinrich Dumoulin to Thomas Merton, September 19, 1964; unpublished letter in the Thomas Merton Studies Center.
12. *Zen and the Birds of Appetite*, pp. 9, 99; Thomas Merton to Father Jean Daniélou, April 21, 1960; unpublished letter in the Thomas Merton Studies Center.
13. *Zen and the Birds of Appetite*, p. 61.
14. Dumoulin, *A History of Zen Buddhism*, pp. 35-36, 126-32.
15. Thomas Merton to Dom Jean Leclercq, OSB, July 23, 1963; unpublished letter in the Thomas Merton Studies Center.
16. 'Learning to Live,' July, 1967 *Thomas Merton's Collected Essays* (unpublished collection in Thomas Merton Studies Center), vol. 12, p. 475. [Published in *Love and Living*, 1979 –ed.]
17. *Ibid.*, p. 475.
18. *Ibid.*, p. 475.
19. *Ibid.*, p. 476.
20. *Mystics and Zen Masters*, p. 10.
21. Thomas Hoover, *The Zen Experience* (New York: New American Library, 1980), p. 110.
22. William Blake, *The Portable Blake* (New York: Viking, 1953), p. 150.
23. *Zen and the Birds of Appetite*, p. 48.

24. Elena Malits, CSC, *The Solitary Explorer: Thomas Merton's Transforming Journey* (San Francisco: Harper & Row, 1980), pp. 142-43.
25. *Ibid.,* pp. 38, 103; Monica Furlong, *Merton: A Biography* (San Francisco: Harper & Row, 1980), pp. 227-28.
26. *Mystics and Zen Masters,* pp. 221-24.
27. Dumoulin, *A History of Zen Buddhism,* pp. 81-82.
28. *Ibid.,* p. 82.
29. *Mystics and Zen Masters,* pp. 31-32.
30. *Ibid.,* p. 221.
31. Chalmers MacCormick, 'The Zen Catholicism of Thomas Merton,' *Journal of Ecumenical Studies* 9, No. 4 (1972), 802-818.
32. *Zen and the Birds of Appetite,* p. 45.
33. Heinrich Dumoulin, *Christianity meets Buddhism* trans. John C. Maraldo (LaSalle, Illinois: Open Court, 1974), pp. 121-22.
34. *Zen and the Birds of Appetite,* p. 86.
35. *Ibid.,* pp. 83, 119-29.
36. *Mystics and Zen Masters,* p. 228.
37. *Zen and the Birds of Appetite,* p. 84.
38. *Ibid.,* p. 132.
39. *Ibid.,* p. 134.
40. *Ibid.,* p. 45.
41. *Ibid.,* p. 39.
42. William Johnston, *The Inner Eye of Love: Mysticism and Religion* (San Francisco: Harper & Row, 1978), p. 52.
43. *Ibid.,* p. 66.
44. *Ibid.,* p. 66.
45. Dumoulin, *Christianity meets Buddhism,* pp. 40-42.
46. Daniel J. Adams, 'Methodology and the Search for a New Spirituality,' *The South East Asia Journal of Theology* 18, No. 2 (1977), 24.
47. Daisetz Teitaro Suzuki, *Mysticism: Christian and Buddhist* (New York: Harper & Brothers, 1957), p. 3-35.
48. Wu, *The Golden Age of Zen,* p. 10.
49. *Ibid.,* p. 11.
50. Sister Therese Lentfoehr, *Words and Silence: On the Poetry of Thomas Merton* (New York: New Directions, 1979), p. 55.
51. *Ibid.,* p. 57.
52. Raymond Bernard Blakney, trans., *Meister Eckhart: A Modern Translation* (New York: Harper & Row, 1941), p. 231; *Zen and the Birds of Appetite,* p. 9.
53. *Zen and the Birds of Appetite,* p. 8.

54. *Ibid.*, p. 11; Blakney, *Meister Eckhart,* p. 205.
55. *Zen and the Birds of Appetite,* p. 5.
56. Thomas Merton, *Cables to the Ace: or Familiar Liturgies of Misunderstanding* (New York: New Directions, 1968), p. 58.
57. *Zen and the Birds of Appetite,* pp. 110-111.
58. *Ibid.*, p. 8.
59. Dumoulin, *Christianity meets Buddhism,* pp. 179-81.
60. Johnston, *The Inner Eye of Love,* p. 117.
61. *Ibid.*, p. 117.
62. Franz Pfeiffer, *Meister Eckhart,* trans. C. de B. Evans (London: John M. Watkins, 1947), I, 147.
63. *Ibid.*, p. 41.
64. *Zen and the Birds of Appetite,* pp. 58, 101-03.
65. *Ibid.*, pp. 22-23.
66. Thomas Merton, *Faith and Violence: Christian Teaching and Christian Practice* (Notre Dame: Univ. of Notre Dame Press, 1968), pp. 259-87.
67. Wu, *The Golden Age of Zen,* p. 15.
68. *Mystics and Zen Masters,* p. 300.
69. *Zen and the Birds of Appetite,* p. 101.
70. Thomas Merton to Marco Pallis, November 14, 1965; unpublished letter in the Thomas Merton Studies Center.
71. *Mystics and Zen Masters,* p. 254.
72. G. B. Sansom, *Japan: A Short Cultural History* rev. ed. (New York: Appleton-Century-Crofts, 1962), p. 354; William Barrett, ed., *Zen Buddhism: Selected Writings of D. T. Suzuki* (Garden City: Doubleday Anchor, 1956), pp. 288-89.
73. Dumoulin, *A History of Zen Buddhism,* p. 228.

Chapter Three

1. *Opening the Bible,* p. 51.
2. *Ibid.*
3. *Ibid.*
4. Huxley, *Ends and Means,* pp. 268-272.
5. Brahmachari is probably an adopted name. It means celibate student and is applied to Hindu monks in general.
6. Benoy Gopal Ray, *Religious Movements in Modern Bengal* (Santiniketan: Visva-Bharati 1965), pp. 94-96.
7. *Seven Storey Mountain,* pp. 193-94.
8. *Ibid.*, p. 193-94.

9. *Ibid.*, p. 157.
10. *Ibid.*, p. 198.
11. 'Dr. M. D. Brahmachari: A personal tribute,' *Thomas Merton's Collected Essays* (unpublished collection in Thomas Merton Studies Center), vol. 9, p. 29.
12. *Ibid.*, p. 28.
13. Vishvanath S. Navarane, *Ananda K. Coomaraswamy* (Boston: Twayne, 1978), p. 10.
14. Ananda K. Coomaraswamy, *Am I My Brother's Keeper?* (Free Port, N.Y.: Books for Libraries Press, 1947), p. 10.
15. Ananda K. Coomaraswamy, *The Transformation of Nature in Art* (New York: Dover, 1956), p. 3.
16. Correspondence of Thomas Merton with Mrs. A. K. Coomaraswamy; unpublished; Thomas Merton Studies Center.
17. Ananda K. Coomaraswamy, *Hinduism and Buddhism* (New York: Philosophical Library, n.d.), p. 27.
18. Thomas Merton to Mrs. A. K. Coomaraswamy, December 18, 1963; unpublished; Thomas Merton Studies Center.
19. *Emblems of a Season of Fury,* p. 78.
20. Thomas Merton, 'Nature and Art in William Blake' (M.A. Thesis, Columbia University, 1939; published as Appendix I in *The Literary Essays of Thomas Merton* (New York: New Directions, 1981); *Conjectures of a Guilty Bystander,* pp. 25, 188; Thomas Merton, ed., *Gandhi on Non-Violence* (New Directions, 1964), pp. 2, 3, 17.
21. *Seven Story Mountain,* pp. 85-87.
22. *Ibid.*, p. 88.
23. *Ibid.*, p. 189.
24. 'Nature and Art in William Blake' (thesis, p. 61).
25. *Ibid.*, p. 48.
26. A. L. Basham, *A Cultural History of India* (Oxford: Clarendon, 1975), pp. 201, 313, 426-29.
27. 'Nature and Art in William Blake' (thesis, p. 63).
28. *Ibid.*, p. 63; Coomaraswamy, *The Transformation of Nature in Art,* p. 11.
29. *The Sign of Jonas* p. 243; Edward Rice, *The Man in the Sycamore Tree* (Garden City: Doubleday Image, 1972), p. 101.
30. *The Ascent to Truth,* p. 26.
31. 'The Inner Experience: Notes on Contemplation,' (fourth draft, p. 29), published in *Cistercian Studies,* 1983.
32. Holographic Journal #52, undated Working Notebook; Swami Prabhavananda and Christopher Isherwood, trans., *How to Know God: The Yoga Aphorisms of Patanjali* (New York: Harper and Brothers, 1953), pp. 29, 54-65, 97. The eight steps of Patanjali

are: abstention from evil doing (*yama*), the various observances (*niyama*), posture (*asana*), control of vital energy (*pranayama*), withdrawal of the senses or interiorization (*pratyahara*), concentration (*dharana*), meditation (*dhyana*), and absorption or union (*samadhi*).

33. *The Inner Experience* (draft pp. 28-30).
34. Thomas Merton, 'Yoga,' Tape 304 A, The Thomas Merton Tapes, ed., Norm Kramer (Chappaqua, New York: Electronic Paperbacks, 1972).
35. Basham, *The Wonder that was India,* pp. 158-159.
36. 'Renewal and Discipline in the Monastic Life,' *Cistercian Studies* 5 (1970), 3-18.
37. Dom Denys Rutledge, *In Search of a Yogi* (London: Routledge & Kegan Paul, 1962), p. viii; Jean-Marie Dechanet, *Yoga Chrétien en dix leçons* (Bruges: Desclee de Brower, 1964).
38. 'Preface' to Dom Denys Rutledge, *In Search of a Yoga,* p. xi.
39. 'Moines et Spirituels Non-Chrétiens,' Extraits des *Collectanea Cisterciensis* (January-March, 1965), 80.
40. Thomas Merton to Arthur Osborne, May 8, 1965; unpublished letter in the Thomas Merton Studies Center.
41. 'Review of the collected works of Ramana Maharshi,' *Collectanea Cisterciensis* (January-March, 1965), 80.
42. *The Inner Experience* (draft pp. 2-5).
43. Abhishiktananda, *Saccidananda* (Delhi: I.S.P.C.K., 1974), p. 71; see also 'Fr. Bede Griffiths–Swami Amaldas here for Monastic Lecture Tour–July 1,' *North American Board for East-West Dialog* No. 5 (May, 1979), 1; Raymond Panikkar, 'Advaita and Bhakti. Love and Identity in a Hindu-Christian Dialogue,' *Journal of Ecumenical Studies* 7, No. 2 (Spring, 1970), 299-309; Naomi Burton, Brother Patrick Hart and James Laughlin, eds., *The Asian Journal of Thomas Merton* (New York: New Directions, 1973), p. 75. For a biography of Monachanin, less well known to anglophones than to French speakers, see *In Quest of the Absolute* by J. G. Weber (Cistercian Publications, 1977).
44. *The Ascent to Truth,* p. 147; Thomas Merton, 'Gurus and Jesus,' Tape 272.
45. Tape 272.
46. *The Inner Experience* (draft p. 28).
47. Vinoba, *Talks on the Gita* (Banares: Sarva Seva Sangh, 1970); *The Asian Journal,* p. 353.
48. *The Inner Experience* (draft p. 29).
49. *The Asian Journal,* p. 351.

50. *Ibid.*, p. 351.
51. *The Inner Experience* (draft p. 29); the verses from the Gita were copied by Merton from Sir Edwin Arnold, trans., *The Song Celestial or Bhagavad-Gita* (London: Routledge & Kegan Paul, 1967), p. 36. Arnold's rendition was Gandhi's favorite version of the Gita, see Gandhi, *An Autobiography*, p. 67.
52. *New Seeds of Contemplation*, pp. 290-97.
53. Tape 304A.
54. S. M. Pathak, *American Missionaries and Hinduism* (Delhi: Manoharlal, 1967), pp. 120, 124; Tape 304A.
55. Thomas Merton, 'Discussion of the Yogi,' Tape 110 A.
56. Tape 110a.
57. Rajamanickam, S. J., *The First Oriental Scholar* (Tirunelveli: De Nobili Research Institute, 1972), pp. 19, 54-57, 59).
58. *Seeds of Destruction*, pp. 159, 163; *Conjectures of a Guilty Bystander* pp. 203-05; *Mystics & Zen Masters*, p. 85.
59. *In Search of a Yogi*, pp. vii-xii.
60. Letter to Philip L. Griggs (Swami Yogeshananda), June 22, 1965; unpublished letter in the Thomas Merton Studies Center.
61. *Ibid.*
62. *Ibid.*
63. *Dr. M. B. Brahmachari*, p. 27.
64. *Gandhi on Non-Violence*, p. 4.
65. *Ibid.*, p. 7.
66. *Ibid.*, p. 8; *Seeds of Destruction*, p. 159.
67. *Ibid.*, pp. 12-15.
68. Ananda K. Coomaraswamy, 'On Being in One's Right Mind,' *Review of Religions*, 7 (November, 1942), 34; *Faith and Violence*, pp. 143-44.
69. *Gandhi on Non-Violence*, p. 20.
70. *Ibid.*, p. 20.
71. *The Asian Journal*, p. 131; Rice, *The Man in the Sycamore Tree*, p. 175.
72. Rice, *The Man in the Sycamore Tree*, p. 175.
73. *The Asian Journal*, p. 195.
74. Basham, *The Wonder that was India*, pp. 304-05; Terence Connolly, trans. *St. Bernard on the Love of God* (New York: Spiritual Book Associates, 1937); Kieran Kavanaugh and Otilio Rodriguez, trans. *The Collected Works of St. John of the Cross* (Washington, D.C.: Institute of Carmelite Studies, 1973).
75. *Ibid.*, p. 332.
76. Rice, *The Man in the Sycamore Tree*, pp. 174-75.

Chapter Four

1. Letter by Father Heinrich Dumoulin to Thomas Merton, September 16, 1967; unpublished letter in the Thomas Merton Studies Center.
2. Petition by Thomas Merton, September 24, 1964; unpublished letter in the Thomas Merton Studies Center.
3. Thomas Merton to Father Dumoulin, November 29, 1964; unpublished letter in the Thomas Merton Studies Center.
4. *Ibid.*
5. An international organization with headquarters in Washington, D.C., founded in 1960, for encouraging cooperation among the world religions.
6. *The Asian Journal,* p. 295.
7. Hart, *Thomas Merton, Monk: A Monastic Tribute,* p. 203.
8. *The Asian Journal,* p. 313.
9. *Ibid.,* p. 296.
10. *Ibid.,* p. 4.
11. Thomas Merton to Seymour Freedgood, April 12, 1967; unpublisehd letter in the Thomas Merton Studies Center.
12. *The Asian Journal,* p. 4.
13. *Zen and the Birds of Appetite,* p. 86.
14. Hart, *Thomas Merton, Monk: A Monastic Tribute,* p. 220.
15. *The Asian Journal,* pp. 4-5.
16. *Ibid.,* p. 8.
17. *Ibid.,* pp. 10, 24, 37, 70, 213, 225, 253.
18. *Ibid.,* pp. 4, 248.
19. *Ibid.,* pp. 147, 151, 161, 183, 197, 212, 238, 245.
20. *Ibid.,* p. 297.
21. *Ibid.,* p. 28.
22. *Ibid.,* p. 65.
23. *Ibid.,* p. 143.
24. *Ibid.,* p. 143.
25. *Ibid.,* p. 144.
26. *Ibid.,* p. 144.
27. *Ibid.,* p. 82.
28. *Ibid.,* p. 125.
29. *Ibid.,* p. 94.
30. *Ibid.,* pp. 179, 274.
31. *Ibid.,* p. 59.

32. T. R. Murti, *The Central Philosophy of Buddhism: A Study of the Madhyamika System* (London: G. Allen and Unwin, 1955), p. 228.
33. *The Asian Journal*, p. 179.
34. *Ibid.*, p. 252.
35. *Ibid.*, p. 10.
36. *Ibid.*, p. 25.
37. *Ibid.*, p. 150.
38. *Ibid.*, p. 28.
39. *Ibid.*, p. 171.
40. *Ibid.*, pp. 78-79; 192-98, 321.
41. K. A. Nilakanta Sastri, *A History of South India: From Prehistoric Times to the Fall of Vijayanagar* (London: Oxford University Press, 1958), pp. 144-45.
42. *The Asian Journal*, pp. 233-36.
43. *Ibid.*, p. 310.
44. *Ibid.*, p. 312.
45. Johnston, *The Inner Eye of Love*, p. 16; *The Asian Journal*, p. 314.
46. *The Asian Journal*, p. 315.
47. *Ibid.*, p. 315.
48. *Ibid.*, p. 315.
49. *Ibid.*, p. 316.
50. *Ibid.*, p. 317.
51. *Ibid.*, p. 313.
52. *Ibid.*, p. 338.
53. *The Asian Journal*, pp. 340-41.
54. *Ibid.*, p. 119.
55. *Ibid.*, p. 125.
56. *Ibid.*, p. 327.
57. *Ibid.*, p. 327.
58. *Ibid.*, p. 329.
59. *Ibid.*, p. 334.
60. *Ibid.*
61. *Ibid.*, p. 342.
62. *Ibid.*
63. Walter Nigg, *Vom beispielhaften Leben: Neun Leitbilder und Wegweisungen* (Olten and Freiburg: Walter-Verlag, 1974), pp. 207-24.
64. *The Asian Journal*, p. 343.

Epilogue

1. Furlong, p. 287.
2. *The Asian Journal,* pp. 103, 138.
3. *Ibid.,* p. 238.
4. Austin Flannery, O.P., gen. ed., *Vatican Council II: The Conciliar and Post Conciliar Documents* (Northport, New York: Costello, 1975), p. 738.
5. *Emblems of a Season of Fury,* p. 383.
6. *Dr. M. B. Brahmachari,* p. 27.

BIBLIOGRAPHY

BOOKS BY MERTON

The Ascent to Truth.
New York: Harcourt, Brace, 1951.
The Asian Journal of Thomas Merton.
Eds. Naomi Burton, Brother Patrick Hart, and James Laughlin. Consulting Editor Amiya Chakravarty. New York: New Directions, 1973.
Cables to the Ace: or Familiar Liturgies of Misunderstanding.
New York: New Directions, 1968.
The Conjectures of a Guilty Bystander.
Garden City: Doubleday, 1966.
Contemplation in a World of Action.
Garden City: Doubleday, 1971.
Emblems of a Season of Fury.
New York: New Directions, 1963.
Faith and Violence: Christian Teaching and Christian Practice.
Notre Dame: University of Notre Dame Press, 1968.
Gandhi on Non-Violence: Selected Texts from Mohandas K. Gandhi's 'Non-Violence in Peace and War'.
Ed. with an Intro. by Thomas Merton. New York: New Directions, 1965.
Mystics and Zen Masters,
1967; rpt. New York: Dell, 1978.
New Seeds of Contemplation.
New York: New Directions, 1972.
Opening the Bible.
Collegeville, MN: Liturgical Press, 1972. London: G. Allen and Unwin, 1972.
Seeds of Destruction.
New York: Macmillan, 1967.
The Seven Storey Mountain.
New York: Harcourt, Brace, 1948.

The Sign of Jonas.
New York: Harcourt, Brace, 1953.
A Thomas Merton Reader.
Edited by Thomas P. McDonnell. Rev. ed. Garden City: Doubleday, 1974.
The Waters of Siloe.
New York: Harcourt, Brace, 1949.
The Way of Chuang Tzu.
New York: New Directions, 1965.
The Wisdom of the Desert: Sayings from the Desert Fathers of the Fourth Century.
Translated by Thomas Merton. New York: New Directions, 1960.
Zen and the Birds of Appetite.
New York: New Directions, 1968.

INTRODUCTION, REVIEWS AND PREFACES BY MERTON

'A Christian Looks at Zen'.
Introduction to John C. H. Wu, *The Golden Age of Zen.* Taipei, Taiwan: The National War College, 1967.
'Moines et Spirituels Non-Chrétiens'.
Extraits des *Collectanea Cisterciensia* (January-March, 1965), 80.
'Review of the Collected Works of Ramana Maharshi'.
Collectanea Cisterciensia (January-March, 1965), 80.
'Author's Preface to the Japanese Edition'.
Nanae No Yama [The Seven Storey Mountain], trans. Tadishi Judo. Tokyo: Chuo Shupansha, 1966, pp. 11-12.

TAPES BY MERTON

'Discussion of the Yogi'. Tape 110A.
The Thomas Merton Tapes, ed. Norm Kramer. Chappaqua, New York: Electronic Paperbacks, 1972.
'Gurus and Jesus'. Tape 272.
The Thomas Merton Tapes, ed. Norm Kramer, Chappaqua, New York: Electronic Paperbacks, 1972.
'Yoga'. Tape 304A.
The Thomas Merton Tapes, ed. Norm Kramer. Chappaqua, New York: Electronic Paperbacks, 1972.

UNPUBLISHED MATERIALS

'Dr. M. B. Brahmachari. A personal tribute'.
Thomas Merton's Collected Essays, Unpublished collection in Thomas Merton Studies Center, vol. 9.
Holography Journal #52. Undated Working Notebook.
'The Inner Experience: Notes on Contemplation'. (Fourth Draft) [to be published serially in the journal *Cistercian Studies* 1983–1984–ed.]
'Learning to Live'. July, 1967. *Thomas Merton's Collected Essays.* Unpublished collection in Thomas Merton Studies Center, vol. 12. [Published in 1979 in *Love and Living* –ed.]
'Nature and Art in William Blake'. Unpublished M.A. Thesis, Columbia University, 1939. [Published in 1981 in *The Literary Essays of Thomas Merton* –ed.]
'Renewal and Discipline in the Monastic Life'. *Thomas Merton's Collected Essays.* Unpublished collection in Thomas Merton's Studies Center. [Published in *The Literary Essays of Thomas Merton* (Brother Patrick Hart, editor), Appendix I. New York: New Directions, 1981–ed.]

BOOKS ABOUT MERTON

Adams, Daniel J. *Thomas Merton's Shared Contemplation: A Protestant Perspective,* Kalamazoo, Michigan: Cistercian Publications, 1979.
Furlong, Monica. *Merton, A Biography.* San Francisco: Harper & Row, 1980.
Hart, Brother Patrick, ed. *Thomas Merton, Monk: A Monastic Tribute.* Garden City: Doubleday, 1976.
Hart, Brother Patrick, ed. *The Message of Thomas Merton.* Kalamazoo: Cistercian Publications, 1981.
Lentfoehr, Sister Thérèse. *Words and Silence: On the Poetry of Thomas Merton.* New York: New Directions, 1979.
Malits, Elena, CSC. *The Solitary Explorer: Thomas Merton's Transforming Journey.* San Francisco: Harper & Row, 1980.
Rice, Edward. *The Man in the Sycamore Tree: The Good Times and Hard Life of Thomas Merton.* Garden City: Doubleday, 1972.

ARTICLE ABOUT MERTON

MacCormick, Chalmers. 'The Zen Catholicism of Thomas Merton'. *Journal of Ecumenical Studies* 9 (1972) 802-18.

WORKS OF RELATED INTEREST

BOOKS

Abhishiktanada, Saccidananda. *A Christian Approach to Advaitic Experience.* Delhi: I.S.P.C.K., 1974.

Arnold, Sir Edwin, trans. *The Song Celestial or Bhagavad-Gita.* London: Routledge & Kegan Paul, 1967.

Barrett, William, ed. *Zen Buddhism: Selected Writings of D.T. Suzuki.* Garden City: Doubleday, 1956.

Basham, A. L. *A Cultural History of India.* Oxford: Clarendon Press, 1975.

Basham, A. L. *The Wonder that was India.* New York: Grove Press, 1959.

Blake, William. *The Portable Blake.* New York: Viking Press, 1953.

Blakney, Raymond Bernard, trans. *Meister Eckhart: A Modern Translation.* New York: Harper & Row, 1941.

Blofeld, John, trans. *The Zen Teachings of Huang Po on the Transmission of Mind.* London: Rider, 1958.

Ching, Julia. *Confucianism and Christianity: A Comparative Study.* Tokyo: Kodansha International/Sophia University, 1977.

Connolly, Terence, trans. *St. Bernard on the Love of God* New York: Spiritual Book Associates, 1937.

Coomaraswamy, Ananda K. *Am I my Brother's Keeper?* Freeport, New York: Books for Libraries Press, 1947.

Coomaraswamy, Ananda K. *The Dance of Shiva.* New York: Noonday Press, 1957.

Coomaraswamy, Ananda K. *Hinduism and Buddhism.* New York: Philosophical Library [1943].

Coomaraswamy, Ananda K. *The Transformation of Nature in Art.* New York: Dover, 1956.

Déchanet, Jean-Marie. *Yoga Chrétien en diz Leçons.* Bruges: Desclée de Brower, 1964.

Dumoulin, Heinrich. *A History of Zen Buddhism.* Boston: Beacon

Press, 1963.

Dumoulin, Heinrich. *Christianity meets Buddhism.* Trans. John C. Maraldo. Lasalle, Illinois: Open Court, 1974.

Flannery, Austin OP, gen. ed. *Vatican Council II: The Conciliar and Post Conciliar Documents.* Northport, New York: Costello, 1975.

Gandhi, Mohandas K. *An Autobiography: The Story of My Experiments with Truth.* Boston: Beacon Press, 1957.

Graham, Aelred. *Zen Catholicism.* New York: Harcourt, Brace and World, 1963.

Hoover, Thomas. *The Zen Experience.* New York: New American Library, 1980.

Huxley, Aldous. *Ends and Means: An Inquiry into the Nature of Ideals and in the Method Employed for Their Realization.* New York: Harper, 1937.

Johnston, William. *The Inner Eye of Love: Mysticism and Religion.* San Francisco: Harper & Row, 1978.

Kavanaugh, Kieran and Rodriguez, Otilio, trans. *The Collected Works of St. John of the Cross* Washington, D.C.: Institute of Carmelite Studies, 1973.

Luk, Charles, ed. and trans. *The Transmission of the Mind outside the Teaching.* New York: Grove Press, 1974.

Moffit, John. *Journey to Gorakhpur.* New York: Holt, Rinehart and Winston, 1972.

Murti, T. R. V. *The Central Philosophy of Buddhism: A Study of the Madhyamika System.* London: G. Allen and Unwin, 1955.

Naravane, Vishwanath S. *Ananda K. Coomaraswamy.* Boston: Twayne, 1978.

Nigg, Walter. *Vom beispielhaften Leben: Neun Leitbilder und Wegweisungen.* Olten and Fribourg: Walter-Verlag, 1974.

Pathak, S. M. *American Missionaries and Hinduism.* Delhi: Manoharlal, 1967.

Pfeiffer, Franz. *Meister Eckhart.* Trans. C. de B. Evans. 2 vols. London: John M. Watkins, 1931-47.

Prabhavananda, Swami and Christopher Isherwood, trans. *How to Know God: Yoga Aphorisms of Patanjali.* New York: Harper and Brothers, 1953.

Rajamanickam, S. *The First Oriental Scholar.* Tirunelveli: De Nobili Research Institute, 1972.

Ray, Benoy Gopal. *Religious Movements in modern Bengal.* Santiniketan: Visva-Bharati, 1965.

Rutledge, Dom Denys. *In Search of a Yogi.* London: Routledge and Kegan Paul, 1962.

Sansom, G. B. *Japan: A Short Cultural History.* New York: Appleton-Century-Crofts, 1962.

Sastri, K. A. Nilakanta. *A History of South India: From Prehistoric Times to the Fall of Vijayanagar.* London: Oxford University Press, 1958.

Suzuki, Daisetz Teitaro. *Mysticism: Christian and Buddhist.* New York: Harper & Brothers, 1957.

Vinoba. *Talks on the Gita.* Banares: Sarva Seva Sangh, 1970.

Wu, John C. H. *Chinese Humanism and Christian Spirituality.* Jamaica, New York: St. John's University Press, 1965.

Wu, John C. H. *The Golden Age of Zen.* Taipei, Taiwan: The National War College, 1967.

ARTICLES

Adams, Daniel J. 'Methodology and the Search for a New Spirituality', *The South East Asia Journal of Theology* 18, No. 2 (1977) 13-25.

'Fr. Bede Griffiths–Swami Amaldas here for Monastic Lecture Tour–July 1'. *North American Board for East-West Dialog* No. 5 (May, 1979), 1.

Coomaraswamy, Ananda K. 'On Being in One's Right Mind'. *Review of Religions* 7 (November, 1942), 32-39.

Panikkar, Raymond. 'Advaita and Bhakti. Love and Identity in a Hindu-Christian Dialogue'. *Journal of Ecumenical Studies* 7, No. 2 (Spring, 1970), 299-309.

CISTERCIAN PUBLICATIONS INC.

TITLES LISTING

THE CISTERCIAN FATHERS SERIES

THE WORKS OF BERNARD OF CLAIRVAUX

Treatises I: Apologia to Abbot William, On Precept and Dispensation . . . . CF 1

On the Song of Songs I–IV . . . . CF 4,7,31,40

The Life and Death of Saint Malachy the Irishman . . . . CF 10

Treatises II: The Steps of Humility, On Loving God . . . . CF 13*

Magnificat: Homilies in Praise of the Blessed Virgin Mary [with Amadeus of Lausanne] . . . . CF 18

Treatises III: On Grace and Free Choice, In Praise of the New Knighthood . . . . CF 19

Sermons on Conversion: A Sermon to Clerics, Lenten Sermons on Psalm 91 . . . . CF 25

Five Books on Consideration: Advice to a Pope . . . . CF 37

THE WORKS OF WILLIAM OF SAINT THIERRY

On Contemplating God, Prayer, and Meditations . . . . CF 3*

Exposition on the Song of Songs . . . . CF 6

The Enigma of Faith . . . . CF 9

The Golden Epistle . . . . CF 12

The Mirror of Faith . . . . CF 15

Exposition on the Epistle to the Romans . . . . CF 27

The Nature and Dignity of Love . . . . CF 30

THE WORKS OF AELRED OF RIEVAULX

Treatises I: On Jesus at the Age of Twelve, Rule for a Recluse, The Pastoral Prayer . . . . CF 2

Spiritual Friendship . . . . CF 5

The Mirror of Charity . . . . CF 17†

Dialogue on the Soul . . . . CF 22

THE WORKS OF GILBERT OF HOYLAND

Sermons on the Song of Songs I–III . . CF 14,20,26

Treatises, Sermons, and Epistles . . . . CF 34

THE WORKS OF JOHN OF FORD

Sermons on the Final Verses of the Song of Songs I–VI . . . . CF 29,39,43,44,45,46

OTHER EARLY CISTERCIAN WRITERS

The Letters of Adam of Perseigne, I . . . . CF 21

Alan of Lille: The Art of Preaching . . . . CF 23

Idung of Prüfening. Cistercians and Cluniacs: The Case for Cîteaux . . . . CF 33

The Way of Love . . . . CF 16

Guerric of Igny. Liturgical Sermons I–II . . . . CF 8,32

Three Treatises on Man: A Cistercian Anthropology . . . . CF 24

Isaac of Stella. Sermons on the Christian Year, I . . . . CF 11

Stephen of Lexington. Letters from Ireland, 1228–9 . . . . CF 28

THE CISTERCIAN STUDIES SERIES

MONASTIC TEXTS

Evagrius Ponticus. Praktikos and Chapters on Prayer . . . . CS 4

The Rule of the Master . . . . CS 6

The Lives of the Desert Fathers . . . . CS 34

Dorotheos of Gaza. Discourses and Sayings . . . . CS 33

Pachomian Koinonia I–III:
- The Lives . . . . CS 45
- The Chronicles and Rules . . . . CS 46
- The Instructions, Letters, and Other Writings of St Pachomius and His Disciples . . . . CS 47

Besa, The Life of Shenoute . . . . CS 73

** Temporarily out of Print* *† Forthcoming*

CHRISTIAN SPIRITUALITY

MONASTIC STUDIES

CISTERCIAN STUDIES

Studies in Medieval Cistercian History sub-series

** Temporarily out of print* *† Forthcoming*

THOMAS MERTON

The Climate of Monastic Prayer . . . . CS 1
Thomas Merton on St Bernard . . . . CS 9
Thomas Merton's Shared Contemplation: A Protestant Perspective (Daniel J. Adams) . . . . CS 62
Solitude in the Writings of Thomas Merton (Richard Anthony Cashen) . . . . CS 40
The Message of Thomas Merton (Brother Patrick Hart, ed.) . . . . CS 42
Thomas Merton Monk (revised edition/ Brother Patrick Hart, ed.) . . . . CS 52
Thomas Merton and Asia: A Quest for Utopia (Alexander Lipski) . . . . CS 74†

THE CISTERCIAN LITURGICAL DOCUMENTS SERIES †

The Cistercian Hymnal: Introduction and Commentary . . . . CLS 1
The Cistercian Hymnal: Text and Melodies . . . . CLS 2
The Old French Ordinary and Breviary of the Abbey of the Paraclete: Introduction and Commentary . . . . CLS 3
The Old French Ordinary of the Abbey of the Paraclete: Text . . . . CLS 4
The Paraclete Breviary: Text . . . . CLS 5–7
The Hymn Collection of the Abbey of the Paraclete: Introduction and Commentary . . . . CLS 8
The Hymn Collection of the Abbey of the Paraclete: Text . . . . CLS 9

FAIRACRES PRESS, OXFORD

The Wisdom of the Desert Fathers
The Letters of St Antony the Great
The Letters of Ammonas, Successor of St Antony
A Study of Wisdom. Three Tracts by the author of *The Cloud of Unknowing*
The Power of the Name. The Jesus Prayer in Orthodox Spirituality (Kallistos Ware)
Contemporary Monasticism
A Pilgrim's Book of Prayers (Gilbert Shaw)
Theology and Spirituality (Andrew Louth)
Prayer and Holiness (Dumitru Staniloae)

* *Temporarily out of print* † *Forthcoming*